Centering Race in the STEM Education of African American K–12 Learners

This book is part of the Peter Lang Education list.
Every volume is peer reviewed and meets
the highest quality standards for content and production.

PETER LANG
New York • Bern • Berlin
Brussels • Vienna • Oxford • Warsaw

Centering Race in the STEM Education of African American K–12 Learners

Glenda M. Prime, Editor

PETER LANG
New York • Bern • Berlin
Brussels • Vienna • Oxford • Warsaw

Library of Congress Cataloging-in-Publication Data

Names: Prime, Glenda M., editor.
Title: Centering race in the STEM education of African American
K–12 learners / edited by Glenda M. Prime.
Description: New York: Peter Lang, 2019.
Includes bibliographical references.
Identifiers: LCCN 2018047630 | ISBN 978-1-4331-6176-6 (hardback: alk. paper)
ISBN 978-1-4331-6175-9 (paperback: alk. paper) | ISBN 978-1-4331-6177-3 (ebook pdf)
ISBN 978-1-4331-6178-0 (epub) | ISBN 978-1-4331-6179-7 (mobi)
Subjects: LCSH: African Americans—Education. | Science—Study and
Teaching—Social aspects—United States. | Minorities in science—United States.
Discrimination in education—United States.
Culturally relevant pedagogy—United States.
Academic achievement—United States.
Classification: LCC LC2717 .C4 2019 | DDC 371.829/96073—dc23
LC record available at https://lccn.loc.gov/2018047630
DOI 10.3726/b14757

Bibliographic information published by **Die Deutsche Nationalbibliothek**.
Die Deutsche Nationalbibliothek lists this publication in the "Deutsche
Nationalbibliografie"; detailed bibliographic data are available
on the Internet at http://dnb.d-nb.de/.

© 2019 Peter Lang Publishing, Inc., New York
29 Broadway, 18th floor, New York, NY 10006
www.peterlang.com

All rights reserved.
Reprint or reproduction, even partially, in all forms such as microfilm,
xerography, microfiche, microcard, and offset strictly prohibited.

*To children everywhere who strive to learn
against innumerable odds,
and to the teachers who teach them.*

Table of Contents

List of Figures   ix

Acknowledgments   xi

Introduction: Race-Visible Pedagogy in the STEM Education of African American Learners   1
   Glenda Prime

Challenging Whiteness in Science Education   19
   Gale Seiler

Toward a Transformative Framework for STEM Education: Achieving Equity Through a Holistic Approach   35
   Roni Ellington

Reconceptualizing Science Education for Learners of African Descent   71
   Jomo W. Mutegi, Crystal H. Morton, and Leslie K. Etienne

Broadening Millennials' Participation in STEM and the Teaching Professions Through Culturally Relevant, Place-Based, Informal Science Internships 95
    Jacqueline Leonard, Scott A. Chamberlin, Elsa Bailey, Geeta Verma, and Helen Douglass

Developing Pre-Service Mathematics Teachers to Meet the Needs of Black Male Students in Teacher Education Programs 129
    Julius Davis, Ramon B. Goings, and Keisha M. Allen

Toward a Framework for Culturally Relevant Inquiry-Based Science Pedagogy 151
    Vanessa Dodo Seriki

Antiracist Curriculum and Pedagogies in Science Teacher Education 171
    Felicia Moore Mensah

Contributors *189*

Figures

Figure 1.1 Social Context of Classroom Learning   23

Figure 2.1 A Transformative Framework for STEM Education   40

Figure 6.1 *Types of Border Crossings*   155

Figure 6.2 Definition of Inquiry   160

Figure 6.3 Relationship Between CRP and IbSI   163

Figure 6.4 Nested Relationship Between CRP and IbSI   164

Acknowledgments

I gratefully acknowledge the Spencer Foundation for the grant that made possible the conference, titled, "Bridging Research and Practice in K-12 STEM Education for African American Learners." The conference held at Morgan State University in November 2016 was the impetus for this volume.

I am grateful for the technical support provided by Ms. Michelle Baptiste in the preparation of this volume.

Introduction

Race-Visible Pedagogy in the STEM Education of African American Learners

GLENDA PRIME

This volume grew out of a Spencer Foundation-funded conference titled, "Bridging Research and Practice in the STEM Education of African American K-12 Learners." The premise of the conference was that in spite of the substantial volume of research on the issue of STEM education for African American and other minority groups, such students continue to underperform their White counterparts on standardized measures of achievement. The conference proposal argued that the persistent "underachievement" of African American learners with respect to their majority counterparts was due to the fact that the research was fragmented, narrowly disseminated, and to varying extents inaccessible to practitioners, curriculum developers, and policy makers. The conference brought together a select group of nationally recognized researchers whose work addresses the issue of teaching and learning of STEM for African American learners, a cross section of stakeholders, including researchers, K–12 teachers, STEM teacher educators, STEM faculty, and school system administrators to identify the salient findings of this body of research and to discuss their implications for practice and policy. The conference provided an opportunity for this diverse group of stakeholders to hear each other's perspectives at a common forum. What the deliberations actually suggested was that the initial premise about the causes of the persistent achievement patterns needed to be revisited. It became apparent that the research and the reform initiatives intended to address the problem of "underachievement" often derived from a deficit perspective about the achievement patterns of African American learners and that this was perhaps the most telling reason for the failure to achieve more equitable outcomes. In this volume, chapter authors reframe the patterns of achievement for STEM learners in ways that challenge the dominant deficit-based explanations, and explore the theoretical considerations that should inform the design of race-visible pedagogies that have the potential for disrupting the persistent inequities in STEM education outcomes.

We define a race-visible pedagogy as one which explicitly and implicitly addresses issues of race and racism both in the content of the curriculum and in all curriculum processes. It encompasses issues of access, opportunity to learn, and outcomes of learning and is informed by the racialized life experiences of the students for whom it is designed. Together the chapters in this volume begin to trace the contours of a race-visible pedagogy for African American STEM learners. Authors address the issues that are foundational to the centering of race in the STEM education of African American learners by doing the following:

- Exposing the inherently racialized nature of schooling in general, and STEM education in particular.
- Critiquing the deficit perspective from which much of education research, policy, and practice derive.
- Offering an alternative framework for designing STEM education that is transformative rather than reproductive of inequality.
- Providing examples of successful pedagogies and high-achieving African American students in an effort to confront the stereotypical notions about the characteristics of African American learners of STEM.

This volume is unique in its bold advocacy for a race-visible pedagogy for the STEM education of African American learners. The distinctive contribution of this book lies in the fact that it offers both a framework for re-theorizing about the achievement of African American learners, as well as practical approaches to promoting access and success for this population. The theoretical perspectives suggest that a consideration of the origin of schooling for Blacks in America brings race into sharp focus and sheds new light on the so-called achievement gap. When viewed as a historical legacy it becomes clear that the achievement patterns currently experienced by African American learners can only be remediated by a reconceptualization of the goals and processes of STEM education, a reconceptualization in which race is central. The privileging of Whiteness has served to silence conversations about race in mainstream educational spaces and has fostered and maintained the deficit perspectives. This volume is an attempt to add to the voices of those who are attempting to break that silence.

The Historical Roots of African American Achievement Patterns in STEM

Attempts to theorize about current patterns of achievement of African American learners in the United States without consideration of the historical roots of schooling for this population are not likely to yield insightful explanations of those patterns, and even less likely to lead to policies and practices that disrupt those patterns. Dominant explanations of the observed patterns of achievement among African American learners focus on a range of factors from innate and genetically determined deficiencies, lack of parental concern and support, diminished levels of interest in schooling, and a lack of motivation to succeed, to more charitable, though inadequate, explanations such as poorly resourced schools and the existence of a cultural mismatch between schools and the norms, values, speech patterns, and ways of knowing of African American children and their families. The less charitable explanations in the first part of this list are reflective of a deficit perspective of African American children that suggest that they are lacking in the attributes needed for school success. The more charitable ones in the second part of the list suggest that the problem of their "underachievement" could be corrected by upgrading schools and tweaking the curriculum to include relevant cultural referents. None of these explanations takes account of the racism that pervades American society, influences policy decisions, "floats between the lines of the curriculum" (Cochran-Smith, 2000, p. 175), and rears its head in myriads of teacher–student interactions every day. The study of origins holds the greatest potential for making these manifestations of racism visible and allows us to reframe our thinking and propose approaches that might disrupt the persistent inequities in educational outcomes.

Historical accounts of education for Blacks in the late nineteenth century that suggest that provision for their education was motivated by the desire to better the plight of recently freed slaves tell only half of the story. The other half of the story is the federal and state legislation that segregated and excluded the "freedmen" from the benefits of education, and in the face of these exclusions, the determined and persistent efforts of the "freedmen" to educate themselves and their children (Anderson, 1988). The period between 1865 and the first half of the twentieth century was marked by arguments for and against segregation of schools. The arguments for segregation were dominant and though they differed in their specifics from one state to another, all had elements of the idea that it was not good for White children to be educated together with Black children, who were inherently inferior. Tyack (1974) asserts that "[d]uring the nineteenth century no group in the United States had greater faith in the equalizing power of schooling or a clearer understanding of the democratic promise of public education than did Black Americans" (p. 110). However, segregation and the accompanying exclusionary practices persisted, and in some cases were sanctioned, until the ruling in 1954 of the US Supreme Court (Brown v Board of Education of Topeka) that segregated schooling was unjust. That political and legislative actions were ineffective in reversing either the geophysical or educational effects of segregation is evident in the fact that schools in urban America are almost as segregated now as they were in 1954. At the present time three-quarters of all Black and

Latino children attend schools that are 80–100% minority populated, whereas less than 1% of White children attend such schools (Frankenberg, Lee, & Orfield, 2003). Segregated schools and the governmental structures that support them are the result of the deeply entrenched racial hierarchy that was necessary to justify the enslavement of African people. The racial hierarchy is the result of what Mills (2007, p. 15) calls "white ignorance," a cultivated set of false beliefs about Black people that accommodates, and indeed is necessary for, the maintenance of White superiority. Political or legislative actions do nothing to address White ignorance. It is no wonder then that the schooling of Blacks continues to be exclusionary. Indeed, Outlaw(2007) asserts that schools are the principal means by which White ignorance is propagated, and that White as well as "black, brown, yellow, red and mixed" children are being mis-educated to believe themselves to be superior in the case of White children and to be inferior in the case of children of color (p. 197).

Current achievement patterns for African American students in America are the direct legacy of this system of race-based miseducation that has dire social and economic consequences for African American children. In the STEM fields, where success and persistence provide access to the greatest share of society's goods, miseducation is even more obviously a justice issue. Kohli, Pizarro, and Nevarez (2017) provide an analysis of the ways in which contemporary forms of racism continue to impact K–12 schools in America. Drawing on the notion of the "new racism" as advanced by Bonilla-Silva (2006) and Fiske (1993), they suggest that contemporary racism is more subtle and flourishes almost unchallenged because it has become normalized. Several processes contribute to or are the result of the normalization of race in educational spaces. One of these is the diversion of emphasis in the discourse about the achievement patterns of African American learners, from structural and policy analyses to individual and family factors as explanations of these achievement patterns. Structural and policy issues, such as those that restrict access to housing for African American families and result in their being restricted to low-income urban spaces, educated in under-resourced schools with high teacher turnover, and zero-tolerance policies that disproportionally impact African American children, are often ignored in studies of the so-called "achievement gap." In contrast, the focus in the discourse on achievement of individual students and their families is one of the ways of diverting attention from the racialized nature of African American children's learning outcomes in STEM. Such a focus makes African American children and their parents complicit in their own plight, and while it does not deny race-based differences in achievement levels, it gives rise to deficit perspectives and deficit-based solutions that attempt to "fix" what is wrong with such kids. Another approach that serves to normalize racism in schooling is the silencing of conversations about race in classrooms and teacher education programs, a form of racism which Kohli *et al.* (2017) call "evaded racism." In this volume, Gale Seiler (2018) calls attention to the difficulty that conference participants had in maintaining a focus on race even when specifically prompted to do so. Seiler sees the dominance of Whiteness that permeates all aspects of schooling as the source of the silence and the means by which the structures of hegemony are upheld. Perhaps the most effective form of silencing results from the complete absence of any racial referents in the curriculum. Children are mis-educated when the curriculum and pedagogical practices ignore who they are as racial and cultural beings and leave no room for conversations about their relationship to the curriculum content. In the STEM fields the "othering" that this produces is most evident and it is not difficult to see how this is reflected in achievement patterns. The notion of achievement and the achievement gap is itself problematic and derives from the pervading meritocracy with its focus on testing. In this volume, Davis *et al.* draw attention to how the unproblematized acceptance of achievement scores has created the notion of an achievement gap as indicative of the diminished mathematical ability of African American males.

The historical legacy of enslavement and White dominance continues to impact the schooling of African American children in numerous ways. Through the persistence and dominance of institutionalized school and classroom practices, as well as through structural arrangements that limit access, African American children are denied high-quality education. These exclusionary practices and structural arrangements are supported by the pervasive racialized storylines by which such children are characterized and by which Blackness is constructed. These storylines are the products of the "white ignorance" which I argue is a direct legacy of slavery. Understanding the history of African American children's communities and their relationship to STEM produces new understandings and allows us to see past the deficiency explanations that dominate the literature and underlie policy and research on the so-called achievement gaps. In this volume authors uncover some of the ways in which racism impacts the STEM education of African American K–12 learners and propose approaches that challenge the deficit perspective. Collectively these chapters outline the contours of what we call a race-visible pedagogy.

The Case for Race-Visible Pedagogies in STEM Education

We chose the term "race-visible" as a rebuttal of what we perceive to be a reluctance on the part of some researchers, policy makers, and practitioners alike to attribute the inequities in educational outcomes to race and racism and to undertake critical analyses of the impact of racism on schools and schooling. A review of literature reveals that the number of researchers who do so is relatively small. Ladson-Billings, as early as 1998, called for a "structural analysis of racism" and was among those who led the way in the application of critical race theory to education (Ladson-Billings & Tate, 1995). More recently scholars like Martin (2009a) and Mutegi (2011), among others, have been exploring the ways in which race and racism are impacting the STEM education of African American learners and are also explicitly focused on centering the racialized experiences and positionality of African American learners in the curricular experiences of these children. However, there is need for more work in this area to address the persistent inequity in educational outcomes, and calls to specifically foreground race in educational research have come from all sectors of the K–20 spectrum. For example, Harper (2012) makes the argument that unless racism is identified and concretely addressed there is little chance of addressing the racial disparities in higher education. Similarly, researchers at the University of Southern California's School of Education, Center for Urban Education, posit that "The United States will not be able to significantly reduce the disparities in postsecondary attainment, even controlling for income and other factors, without explicit discussion about racial inequity" (p. 5). With reference to the K–12 sector, Kincheloe, Steinberg, Rodriquez, and Chennualt (2000) admonish that educators need to make issues of race central in their classrooms. The term "race-visible" was coined to describe the pedagogies that we seek to advocate and was chosen in preference to terms like "culturally relevant," "anti-racist," and "liberatory" which others have used, and which have many elements in common with our approach, because we wish to be explicit about the centrality of race in all aspects of teaching and learning in STEM. For us the term focuses attention on the *experiences* that we wish African American learners and the teachers who teach them to have as they engage with STEM. The distinction we wish to highlight is that the racialized experiences of African American learners in American society must be taken into account in every aspect of their STEM education, the choice of content, the instructional practices, the teacher/student interactions, the assessment practices all designed in consideration of the experience of Blackness in contemporary American society.

The definition of race that is most influential in informing the approaches described in the ensuing chapters aligns with the work of Lopez (1995) who defines it as "a group loosely bound together by historically contingent, socially significant elements of their morphology and/or ancestry." He further describes race as an "ongoing process of social and political struggle" (p. 193). A race-visible pedagogy is one that brings awareness of that struggle into the classroom and avoids the race-neutral or race-evasive strategies that serve to normalize, make invisible, and perpetuate the dominance of Whiteness. The enactment of race-visible pedagogies is an essentially political act that seeks to equip learners to identify and disrupt the power imbalances that result in inequitable school outcomes. It includes deliberate engagement with the sociohistorical, sociocultural, and economic realities that frame the lives of African American students. Greene and Abt-Perkins

(2003) argue that "we need to engage more fully in the process of making race visible, both as a mark of difference and privilege, and ways that the institutions of school and society have placed minority students in poverty at distinct disadvantages in achieving access to quality education..." (p. 3). Our call for a race-visible pedagogy is based on the assertion of critical race theorists (Bell, 1992; Delgado, 1995; Ladson Billings, 1995) who remind us that race is a permanent, pervasive, and often invisible feature of our society. We argue that the first step in a quest to disrupt the pattern of injustice that racism produces is to make it visible.

What does a race-visible pedagogy in STEM look like? The chapters in this volume address this question by providing both practical examples of race-visible pedagogies and discussions of the theoretical underpinnings that support race-visible pedagogies. The authors of these chapters are all engaged in theorizing about the STEM education of African American learners and in the practice of race-visible pedagogies through their work as teacher educators and researchers. The growing emphasis in STEM fields in the curricula of K–12 schools parallels its increasing importance in the world. Access, persistence, and success in STEM is to a large extent the gateway to the highest paying jobs and to positions of power and prestige, but it is increasingly the case that even lower level jobs of the future will require some degree of technological competence. Apart from the economic consequences of not being competent in STEM, there is the emotional, psychological, and political consequence of exclusion from full participation in a society that is heavily influenced by science and technology. The racialized nature of such exclusion becomes evident when data on STEM achievement are examined.

A number of unfavorable features are characteristic of the conditions under which African American children are taught the STEM disciplines. Science and mathematics have traditionally been taught and practiced without recognition of the culture-ladenness of the epistemological assumptions that underlie those fields. Some researchers have for some time now been pointing this out (Aikenhead, 1996; Bishop, 1989; Jegede & Aikenhead, 1999; Warren & Roseberry, 2011), yet the dominance of White Western ways of knowing has allowed the teaching of science as objective and universal truth to prevail. The knowledges and ways of knowing of communities of color have not been admitted into the traditional science classrooms. Further, the STEM disciplines are resource-intensive, and the system of school funding that is common across all states limits the opportunities for enriching science learning that the predominantly minoritized children who attend school in high-poverty schools can have. The problem of teacher shortages in the STEM fields is greatest in schools in these high-poverty urban school districts. It is also the case that most of the teachers of STEM even in urban districts are White and female. All of this heightens the level of cognitive and emotional challenge that African American learners experience in STEM classrooms, with unfavorable consequences for their success and persistence. What is often not recognized is that these problems are rooted in racist structural school and housing policies, and the deficit views derived from the historic notion of a racial hierarchy are invoked as explanations of the "underachievement" of African American STEM learners. A race-visible pedagogy in STEM engages learners in interrogating the social structures that frame the ways in which they are impacted by and connected to STEM as a mean of developing what Ladson Billings (1995) calls critical consciousness.

The content of the chapters in this volume suggests that the following features should characterize a race-visible pedagogy in STEM for African American K–12 learners.

- Adopts an asset-based view of African American learners and creates opportunities for them to develop and display their identities as competent doers of STEM.
- Foregrounds the past and present contributions of people of color to the enterprise of STEM.
- Helps learners to view Western science as one of several forms of knowledge about the natural world and to view knowledge production, including knowledge production in mathematics and science, as inherently cultural.
- Provides opportunities for learners to bring their own STEM-related knowledges and practices into the STEM classroom.

- Develops awareness and ability to identify and critique the ways in which science and technological applications privilege Whiteness at the expense of African American and other minoritized groups.
- Positions STEM classrooms as counter spaces where conversations about race and racism can be had.

It is clear from these characteristics that what we are proposing has much in common with other pedagogies such as "anti-racist," "liberatory," and "culturally responsive," which terms authors in this volume have referred to. We suggest that the term "race-visible" includes elements of all these, but goes beyond them in that it shifts the focus from the *outcomes* that such approaches are intended to achieve, and places the focus on the kinds of *experiences* that we think would result in more equitable outcomes for African American learners. The characteristics offered above have deep implications for all aspects of the teaching and learning of STEM, including curriculum content, student/teacher interactions, instructional strategies, STEM teacher preparation, and STEM education research.

Curriculum Content

The question here is, "What should be the content of a race-visible pedagogy in STEM?" In Chapter 1 of this volume Seiler suggests that there is need for a reconceptualization of what counts as science content, and Mutegi, Morton, and Etienne in Chapter 3 also call for reconceptualization of the science curriculum so that it serves a socially transformative function. They assert that social transformation requires that African American learners acquire mastery in five areas, one of which is the traditional science content. Theorizing about what content should be included in a race-visible pedagogy for African American K–12 learners requires that we give thought to the role that race has played in the design and implementation of science curricula and be less committed to state-mandated content standards and more to what science enriches the lives of our students. The move toward interdisciplinarity between the traditionally separate STEM disciplines opens the door to inclusion of the sociocultural and political aspects of STEM.

It is often the case that opportunities to address social justice issues lie at the intersections of the STEM disciplines.

Student/Teacher Interactions

Two important factors contribute to the highly racialized nature of STEM classrooms where African American students learn: the fact that African Americans are stereotyped as not having high levels of ability in science, and the fact that the majority of STEM teachers are White. The burden that that places on African American children in STEM classrooms depresses their performance and dampens their interest in STEM. The practice of race-visible pedagogies would require that teachers be intentional about creating classrooms that serve as counter-spaces, that disrupt the prevailing narratives about the ability of Black children in STEM. In this volume, Ellington (Chapter 2) offers a framework for a holistic transformation that is necessary for African American children to achieve success. One component of the framework focuses on developing agency, identity, resiliency, and belonging. Chapters by Seiler (Chapter 1) and Leonard *et al.* (Chapter 4) also make reference to identity formation, and Mutegi *et al.* (Chapter 3) talk about the need to change the nature of the student/teacher relationship to one of "joint productive activity." All of these approaches probably depend more on the psychosocial climate of the STEM classroom than on the curriculum content or instructional strategies. Teachers need to adopt an asset-based mindset that is affirming of their students and allows their brilliance to be displayed. It is as much a way of *being* with children as it is the activities that are engaged in that would result in the positive racial socialization that is necessary for success.

Instructional Strategies

Implied in the chapters of this volume is a radical focus on the learner as an African American learner. The awareness of the situatedness of classroom practice made it advisable that authors resist the idea of presenting a set of best practices to be followed in race-visible classrooms. The Mutegi *et al.* chapter and the Leonard chapter provide empirical data on race-visible strategies employed by the authors. Davis *et al.* in Chapter 5 offer the notion of mathematizing aspects of the ordinary lives of African American children as a way to think about content selection and instructional strategies.

Teacher Preparation

The chapters by Moore Mensah (Chapter 7), Dodo-Seriki (Chapter 6), and Davis (Chapter 5) speak most directly to the preparation of teachers to teach STEM in race-visible ways. Seiler, in Chapter 1, revealed the difficulty of getting conference participants to sustain conversations about race even when specifically prompted to do so. She concluded that Whiteness has been so normalized in American society that it has silenced conversations about race and that teachers do not recognize the need to teach STEM in race-visible ways. The implications of this for teacher preparation and in-service professional development of STEM teachers are huge. Moore Mensah's chapter describes her strategies for giving pre-service teachers the language to engage in conversations about race as a way to confront their own assumptions and to aid in the development of racial literacy. Dodo-Seriki in Chapter 6 creates a framework that combines culturally relevant pedagogy, as advanced by Ladson-Billings (2006) with Inquiry-Based Science Instruction. The framework addresses the need to prepare teachers to meet the needs of African American children, ensuring that both their racialized experiences as African American children and their need to master traditional STEM content are addressed.

Research

It is clear that if we are to disrupt the patterns of inequity that serve to exclude African American children from access to high-quality STEM education and the benefits that derive from such access, research has a major role to play. However, the research itself must be reframed. Much educational research, in STEM as in other curriculum areas, has bought into the dominant deficit perspective of Black children. The most glaring example of that is the fixation on the achievement gap in research on African American children. This focus is particularly problematic because it makes White children the standard by which all children are judged and is thus both the cause and the effect of the deficit thinking. The problem is compounded by the fact that relatively little STEM education research makes race and racism central to the analyses and interpretation, except in cases where comparisons are being made across racial lines. In this volume Davis *et al.* in Chapter 5 draw on Martin's (2009b) framing to suggest the use of the experience lens in the study of learning outcomes, as opposed to research that focuses on achievement. Research that focuses on investigations of the impact of race and racism on the experiences of African American children is likely to reveal that the outcomes of African American children in the STEM disciplines is the result of an opportunity-to-learn gap and say little about their ability in STEM.

Some research questions that seem to be germane to advancing the development of race-visible pedagogies that need to be explored are:

- What learning outcomes are most significant in enabling African American learners to be critically conscious of their own positions and to be able to identify and disrupt the structures and policies that perpetuate their exclusion from STEM?
- How are these outcomes to be measured?
- How is research in STEM impacting the African American children whom we study?
- What are the structural and systemic impediments to the implementation of race-visible pedagogies and how best can these be dismantled?
- How do researchers' own racial identity impact their analyses and interpretation of findings in qualitative studies of STEM education for African American learners?

Concluding Thoughts

Regarding the last of the research questions mentioned above, I accept that as editor of this volume I have a responsibility to be transparent about my own positionality with respect to the issues addressed in this introductory chapter and in the volume as a whole. I am an

Afro-Caribbean woman who migrated to the United States almost 20 years ago, for family reasons, after having been a science educator in the Caribbean for 10 years. Upon my arrival in the United States I assumed the role of coordinator of the Mathematics Education and Science Education doctoral programs at an Historically Black University. My interaction with the students and faculty left me baffled about why, from their perspective, race and racism seemed to be the default explanation for everything that was wrong in science and mathematics education in the K–12 sector, in which most of my students worked either as classroom teachers or as district-level mathematics or science curriculum leaders. Nothing in my prior experiences, either as a student or as a faculty member, had given me anything more than an academic understanding of exclusion or marginalization. I had attended a high-achieving girls' high school and almost everybody who took or taught mathematics or science was Black and female. (There was little more than a sprinkling of privileged White girls in the school). As I began to immerse myself in the K–12 setting in the United States, I spent numerous hours in high school classrooms in an urban school district. I saw firsthand the racial divide between the predominantly African American student body (approximately 80% in that school district) and the predominantly White mathematics and science teachers. I saw numerous examples of disruptive classroom behaviors on the part of students, underteaching of science concepts on the part of teachers, high school boys taller than their teacher doing 4th grade-level busy work, micro-aggressions, and harried teachers struggling valiantly to teach with limited resources. Overall, there were too few occasions of opportunities for deep learning of science concepts. The following vignette from one of my classroom visits stands out in my memory and was a defining moment in my understanding of the problems faced by many African American learners in high school science classrooms.

It was a class of about 20 high school students all of whom were African American. The teacher was a young White woman who had just earned her master's degree in education and was fulfilling her first assignment in this urban school setting. The lesson was on polymers, both synthetic ones in common use and biological ones. The instructional strategy was a well-planned series of timed activities for students, interspersed with mini lectures/explanations and directions given by the teacher. The teacher had just presented a number of polymers, polyvinyl chloride, polystyrene, polyethylene, polysaccharides, among others, and had provided diagrams of the molecular structures of these substances and had given information on the uses of these polymers, when a female student raised her hand and asked, "What does poly mean?" The teacher's response was "Our time is up for this activity, we must move on to the next one." The student's question remained unanswered. This was the routine for the entire class. Each activity had been allotted a particular time and it was stopped when the allotted time had passed.

Without providing the reader with details of the research methods, analyses, and contexts as a warrant for accepting my interpretation of these observations (which it is outside of the scope of this chapter to do), I suggest that a number of aspects of the racialized experiences of African American learners are captured in that vignette. From the perspective of the teacher, time was running out and she needed to "cover" the content. She had a well-structured lesson and she needed to complete it so that everyone had a chance to participate in all of the learning experiences that she had designed. Her demeanor during the class and her conversation with me after the class, suggested that she was also motivated by a custodial concern. There was not going to be any unstructured time in the lesson because those were the times when the classroom might descend into the disruptive behaviors often seen in the school. On the other hand, from the perspective of the student, she had asked a question that was not worthy of the time it would take to answer it and, by extension, she herself was not worthy of the time. In a context in which ability in science is thought to be distributed along racial lines, the teacher's response was confirmation that she had not asked an important question because she, as an African American student, could not do science anyway. It was clear to me that what I had seen were "the messages about identity and difference that float between the lines of the curriculum" (Cochran-Smith, 2000).

Experiences like these and many others with my doctoral students, colleagues, and researchers were the impetus for the Spencer Foundation conference that gave rise to this volume. The volume is essentially about the need for a radical transformation in the way we think about STEM and STEM education for African American learners. It is a call for new asset-based frameworks for understanding and meeting the needs of African American learners.

References

Aikenhead, G. S. (1996). Science education: Border crossing into the subculture of science. *Studies in Science Education, 24*(1), 1–52.

Anderson, J. D. (1988). The education of Blacks in the South, 1860–1935. The University of North Carolina Press.

Bell, D. (1992). *Faces at the bottom of the well*. New York, NY: Basic Book.

Bishop, A. J. (1989). Mathematics education in its cultural context. *Educational Studies in Mathematics, 19*, 179–191.

Bonilla-Silva, E. (2006). *Racism without racists: Colorblind racism and the persistence of racial inequality in the United States*. New York, NY: Rowman and Littlefield.

Center for Urban Education. (2011). *Improving postsecondary attainment. Overcoming challenges to an equity agenda in state policy*. Los Angeles, CA: USC Rossier, School of Education.

Center for Urban Education. (2017). *Overcoming common challenges to an equity agenda in state policy*. Los Angeles, CA: USC Rossier, School of Education.

Cochran-Smith, M. (2000). Blind vision: Unlearning racism in teacher education. *Harvard Educational Review.* 70(2), 157–190.

Delgado, R. (1995). *Critical race theory: The cutting edge*. Philadelphia, PA: Temple University Press.

Fiske, S. T. (1993). Controlling other people: The impact of power on stereotyping. *American Psychologist, 48,* 621–628.

Frankenberg, E., Lee, C., & Orfield , G. (2003). *A multiracial society with segregated schools: Are we losing the dream?* Cambridge, MA: The Civil Rights Project at Harvard University.

Greene, S., & Abt-Perkins, D. (Eds.). (2003). *Making race visible. Literary research for cultural understanding*. New York, NY: Teachers College Press.

Harper, S. R. (2012). Race without racism: How Higher Education researchers minimize racist institutional norms. *The Review of Higher Education, 36*(1), 9–29.

Jegede, O. J., & Aikenhead, G. S. (1999). Transcending cultural borders: Implications for science teaching. *Research in Science and Technological Education, 17*(1), 45–66.

Kincheloe, I., Steinberg, S. R., Rodriquez, N. M., & Chennualt, R. E. (Eds.). (2000). *White reign: Developing whiteness in America.* New York, NY: St. Martin's Press.

Kohli, R., Pizarro, M., & Nevarez, A. (2017). The "new racism" of K-12 schools: Centering critical research on racism. *Review of Educational Research, 41,* 182–202.

Ladson-Billings, G. (2006). Yes, but how do we do it? Practicing culturally relevant pedagogy. In J. Lansman & C. W. Lewis (Eds.), *White teachers/diverse classrooms: A guide to building inclusive schools, promoting high expectations, and eliminating racism* (pp. 29–42). Sterling, VA: Styles.

Ladson-Billings, G., & Tate, W. F. (1995). Toward a critical race theory of education. *Teachers College Record, 97*(1), 47–68.

Lopez, I. (1995). The social construction of race. In R. Delgado (Ed.), *Critical race theory: The cutting edge.* (pp. 479–487). Philadelphia, PA: Temple University Press.

Mills, C. W. (2007). White ignorance. In Sullivan, S., and Tuana, N. (Eds.). *Race and Epistemologies of Ignorance* (pp. 13–38). State University of New York Press.

Martin, D. B. (2009a). Researching race in mathematics education. *Teachers College Record, III,* 295–338.

Martin, D. B. (2009b). *Mathematics teaching, learning and liberation in the lives of Black children.* New York, NY: Routledge.

Mutegi, J. W. (2011). The inadequacies of science for all and the necessity and nature of a socially transformative curricular approach for African American science education. *Journal of Research in Science Teaching, 48,* 301–316.

Outlaw, L. (2007). Social ordering and the systematic production of ignorance. In Sullivan, S., and Tuana, N. (Eds.). *Race and Epistemologies of Ignorance* (pp. 197–212). State University of New York Press.

Tyack, D. B. (1974). *The one best system: A history of American urban education.* Cambridge, MA: Harvard University Press.

Warren, B., & Roseberry, A. S. (2011). Navigating interculturality: African American male students and the science classroom. *Journal of African American Males in Education, 7*(1), 98–114.

Challenging Whiteness in Science Education

GALE SEILER

Abstract
Whiteness is a social construction that works as a tool to maintain the systemic advantage of Whites over other groups. It provides a way of being in the world and a way of responding to interrogations about race that ultimately work to maintain the status quo, cover up institutionalized racism, and silence communities of color. By analyzing discourse from one of the working groups at this Symposium, five ways were identified that whiteness was used to deflect consideration of STEM classrooms as racialized spaces. Though conceptual tools were provided to the participants, with the intention that those tools would facilitate the work of challenging whiteness and grappling with why race matters in science and math education, those tools did not allow the participants to position themselves outside the system that whiteness protects. The propensity to portray goodness in the system and to call upon dominant narratives and counter-examples were powerful ways of denying, evading, subverting, and avoiding the issue of race in STEM education. By revealing and reflecting upon these ways, we may learn how to better challenge this active protection of the status quo and meet the goal of the conference as planned.

Introduction

The recent Symposium held at Morgan State University, from which this edited volume emerged, attempted to place race at the center of the conversation and to learn from STEM education research that uses conceptual and methodological frames that place race at the center of their analysis. The organizers recognized that considerable research has already been undertaken on the characteristics of effective K–12 STEM education for African American students and proposed, through the Symposium, to address possible barriers to implementation of existing research findings and to refocus attention on questions not being asked or constructively addressed. To that end, STEM education researchers were asked to present the most compelling research available and engage participants in conversations around that research with the goal of arriving at new understandings and implications for practice.

In many ways, the Symposium aimed to move us beyond the failures of the dominant system to grapple productively with race in STEM education. This echoes the question raised previously by Jomo Mutegi (2013), one of the Symposium presenters, when he asked why research in science education has not yet provided adequate explanations for why science achievement is a "racially determined phenomenon" (p. 87), that is, for why race matters in science achievement. This chapter will explore how this Symposium and other attempts to refocus attention toward the role that race and racism play in science and mathematics classrooms often elicit responses that ultimately work to maintain the system of racial oppression. While the goal of the Symposium was to create an environment that would support meaningful discussion, learning, and new understandings around race and racism in STEM education, the discussions were thwarted by the normalization and neutrality of whiteness. By exploring and identifying the particular ways that whiteness was used to deflect consideration of science and mathematics classrooms as racialized spaces, we may learn how to better challenge its protection of the status quo.

The Symposium

The participants in the Symposium were all in some way connected to science or mathematics education and presumably committed to improving science and mathematics education for African American students. They included science and math education researchers and faculty members; science and math faculty; middle and high school teachers; school system administrators; and others involved in science and math education, for example, in outreach to schools, informal education, and professional development. The two-day Symposium was organized around three plenary sessions, each followed by a working group session. Participants selected their working group before the Symposium. The working group foci were: classroom practice and pedagogy; teacher preparation; and sociocultural impacts. The five plenary presenters also facilitated the working groups. Prior to the Symposium, each presenter was asked to identify two key readings, which were sent to the participants of their working group in advance. I facilitated the classroom practice and pedagogy working group, which was composed of 12 participants.

My Positionality

My presentation at the Symposium, titled *The role of race and racialized experiences in building identities as science learners*, followed Dr. Mutegi's in the first morning session. I opened by introducing myself and voicing the question that I suspected was on everyone's mind: What could this person, a white woman, have to say about teaching African American students? I noted that I sometimes ask myself that question, and I explained that the core of what I would share comes from what I have learned from students and teachers whom I've known and taught and worked with, and I feel a responsibility to share what has been learned and to see where it can take us.

As I prepared for the Symposium, and in particular, for facilitating the working groups, I had many of the same worries that Sherry Marx (2004) and others have written about, that is, "how to lead my own participants in a productive, critical exploration of whiteness and white racism" (p. 37). Although many of the participants were people of color, I anticipated that whiteness would pervade the conversations, because the day-to-day experiences of all participants are within oppressive systems that privilege whiteness. My goal was to enable my working group participants to move beyond some of the ways that whiteness has shaped their thinking and language. To similar ends, the five presenters, each in different ways, offered new concepts and constructs to facilitate this moving beyond. Considering the working group sessions as discourse communities, the aim was that the working group conversations, which followed the presentations, would be more meaningful and worthwhile because these new concepts would serve as tools to push the participants' thinking beyond where it was (Putnam & Borko, 2000). In the following section, I describe my presentation and the conceptual tools I made available to the participants. For clarity, the names of the tools are underlined as they are introduced.

Presentation on Classroom Practice and Pedagogy

In my presentation, I stressed the important links between learning and *identity construction*, noting that learning is fundamentally linked to social and cultural contexts and processes. I conceptualized identity construction as a negotiation between how you see and position yourself and how others see and position you, referring to the idea that the long-term happens through day-to-day interactions as identity construction continues over time (Holland, Lachiotte, Skinner, & Cain, 1998). The following diagram (Figure 1.1) was used to situate identity construction and learning within a wider social context and to call attention to storylines and narratives that circulate and make available particular resources and subject positions, which shape classroom interactions and, thus, teaching, learning, and identity work in them. The diagram also calls attention to the resources and ways of being that are socially acquired and brought into classrooms by students and teachers. My goal was to encourage the participants to move from thinking about classroom spaces and what happens in them to societal structures and back again.

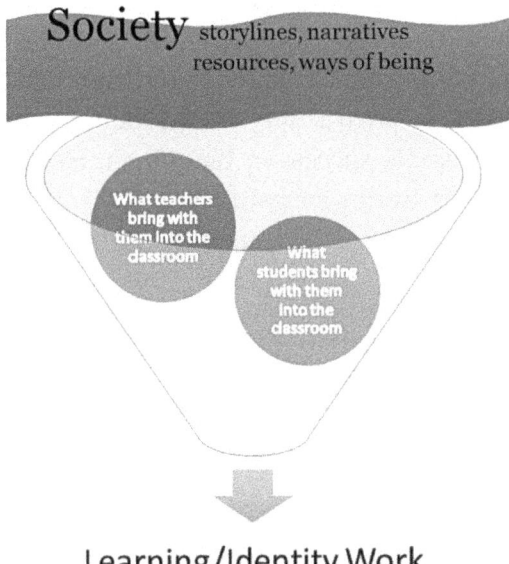

Figure 1.1 Social Context of Classroom Learning (source is author's own).

Racialized Storylines

Drawing from one of the readings sent to the participants in advance, *racial storylines* are pervasive in societal discourse and, therefore, become critical aspects of life in schools (Nasir, Snyder, Shah, & Ross, 2012). Several examples were used in my presentation to illustrate the presence of racialized storylines in our society, both historically and currently. Ava DuVernay's recent film, *Thirteenth*, documents how the construction of blacks as threatening and menacing criminals was a key narrative in rebuilding the South's economy after the Civil War. The film details how these negative depictions have supported practices since then, for example, convict leasing, Jim Crow, and the war on drugs and mass incarceration that we have today.

I drew an example of racialized storylines in education from the experiences of Tiffany Martínez. In her blog (Martínez, 2017b), she described the response of a professor who presumed that her assignment contained plagiarized text. She posted an image of her returned assignment in which the professor questioned whether she could actually have used the word "hence" and suggested that she had copied text from someone else. Martínez described how the professor's comments led her to question her own ability as a student of color in academia, despite her considerable success in that arena. Her blog post elicited numerous responses from other students of color describing similar experiences.

> This morning, my professor handed me back a paper (a literature review) in front of my entire class and exclaimed "this is not your language." On the top of the page they wrote in blue ink: "Please go back and indicate where you cut and paste." They assumed that the work I turned in was not my own. My professor did not ask me if it was my language, instead they immediately blamed me in front of peers. On the second page the professor circled the word "hence" and wrote in between the typed lines "This is not your word." The word "not" was underlined. Twice. My professor assumed someone like me would never use language like that. As I stood in the front of the class while a professor challenged my intelligence I could just imagine them reading my paper in their home thinking *could someone like her write something like this?*
>
> Their blue pen was the catalyst that opened an ocean of self-doubt that I worked so hard to destroy. In front of my peers, I was criticized by a person who had the academic position I aimed to acquire. I am hurting because my professor assumed that the only way I could produce content as good as this was to "cut and paste." I am hurting because for a brief moment I believed them.

I also provided several examples of the presence and power of racialized storylines as part of life in K–12 schools. In Nasir *et al.* (2012), an interview with an African American student illustrates how these negative, racialized storylines impact life in schools, "Because like, the teachers, not all teachers, but some of these teachers, like they expect us to get kicked out, and get kicked out of class, and mess up and get referrals" (p. 290). Even more specifically, another example shows the impact of racialized storylines on participation in class discussions, "Say if, I raise my hand and then, then the Asian person or white person raises his hand, like they would go to the Asian person first and then help the other one later, because they think we gonna say something either stupid, like you know like ooh what is this problem? What page is this on? Or like you know, umm. Like they just will say, they think that Black people will say something out of pocket..." (p. 292).

I shared these examples to illustrate how racialized storylines impose certain positions on African American students (e.g., trouble maker), while other positions are closed to them (e.g., smart student). As we have known for a long time, for example from research on self-fulfilling prophecy and teacher expectations (Van Den Bergh, Denessen, Hornstra, Voeten, & Holland, 2010), such positioning constrains learning as well as identity trajectories over time.

Students' Lives as Resources for Learning and Identity Work

In contrast to the damaging effects of racialized storylines in school and society, research has shown that positive things happen when teachers recognize students' experiences, knowledge, and their ways of being, talking, and sense-making as resources for teaching and learning. The work of Wade Boykin, who also was a Symposium presenter on the following day, and of Luis Moll was used to illustrate this, and the concepts of *cultural dispositions* (Boykin, 1986) and *funds of knowledge* (Moll, Amanti, Neff, & Gonzalez, 1992) were provided as tools to advance participant thinking.

I also provided several examples of how taking an asset view of students necessitates that we see and hear students in new ways. This also requires us to expand how we think about what constitutes science and how science is done, and by doing so, we can begin to counter the pervasive racialized storylines that permeate science classrooms and school walls. I offered to the participants the idea of *classrooms as cultural spaces* as another tool. Examples included Chris Emdin (2010), whose vision of teaching science through *reality pedagogy* is based on hip-hop, and Carol Lee (2001), who uses *cultural modeling* to link students' interpretive skills with rap lyrics to make sense of other literature. These examples show how instruction can be designed to leverage everyday knowledge of youth, especially African American youth, to support discipline-specific learning.

Other examples were used to show critical moments in teachers' decision-making that can either shut down or open up science to students and how these moments can be shaped by racialized storylines (Nasir *et al.*, 2012) that are widespread in society and schools. Using video from my research in Philadelphia and an example from Warren and Rosebery's (2011) work with younger African American students in Boston, the idea of *classrooms as counterspaces* was introduced, where student counternarratives can be constructed to resist and work against racialized narratives imposed on students. The key point of these examples was that there were positive positions available for students, as doers of science and as black youth. Therefore, considering a classroom as a counterspace creates new possibilities for long-term consequences and changes in identity trajectories for young people.

I concluded my presentation by noting that we need to do more than provide access to science learning for African American students; we also need to engage them in a critique of science and in exploring how they might use science to develop critiques of society. In his presentation in the same plenary, Dr. Mutegi drew from the work of Claude Steele on *stereotype threat* (Steele & Aronson, 1995) and used examples from his own research with African American university students to show the power of stereotypes in creating self-doubt and interfering with learning. He also provided several possible defenses against negative stereotypes and racialized storylines, including *positive racial socialization* and *preparation for bias* (Wang & Huguley, 2012).

At the end of my presentation, after summarizing things that we know can make a difference in the science education of African American students, I asked the question: Why don't we do these things? On the final slide, I suggested reasons, several of which I included because I anticipated that they would be on the participants' minds: a focus on standardized tests and accountability; a narrow view of what counts as doing science and mathematics; implicit teacher biases that go unrecognized and unchallenged; the tendency to teach as one has been taught; and the difficulty of preparing teachers to teach in antioppressive ways. After facilitating and listening to the working group conversations that followed each of the three plenary sessions over two days, I recognized another reason that I had not included in the bullet points on the last slide. Although we had offered conceptual tools in the hope of promoting participants to grapple with the question of race in science and math education, there was another tool in use, the tool of whiteness, which protected the status quo and prevented uptake of the new tools.

Whiteness as a Tool

The concept of whiteness describes how white norms permeate society while appearing common and value-neutral. Thus, an aim of critical whiteness studies is to make explicit the ways whiteness and its invisibility lead to the maintenance of white power. Frankenberg (1993) described whiteness as a social construction designed intentionally and purposefully to maintain systemic advantage of whites over other groups. More specifically, Picower (2009) referred to whiteness as a tool, in that it allows a job to be done more effectively, in this case, the job of maintaining "hegemonic stories and dominant ideologies of race, which in turn, uphold structures of White supremacy" (p. 204). Picower further explained that "[i]n an attempt to preserve their hegemonic understandings, participants used these tools to deny, evade, subvert, or avoid the issues raised" (p. 205). Thus, whiteness as a tool (or set of tools) is useful in considering the discussions that took place in the classroom practice and pedagogy working group at the Symposium, and examples from these discussions will be used to illustrate how whiteness provides a way of responding to interrogations about race (in this case, race in STEM education) that ultimately works to maintain the status quo, cover up institutionalized racism, and silence communities of color.

Nonengagement With Conceptual Tools

One of the most striking aspects of the discussions in the working group sessions was the infrequent engagement by participants with concepts from the presentations and readings, which were intended to be tools that could provide a way for participants to challenge and advance their understanding. This lack of engagement with the tools happened even when participants were specifically asked to refer to them or use them, and thus it allowed evasion and avoidance of the issues raised. For example, to kick off the first working group, the participants were asked to think of an experience they had (as learner or teacher) that illustrated one of the ideas from the morning's presentations, for example, racialized narratives, classrooms as counterspaces, stereotype threat, racial socialization, etc. For the most part, in their responses, the participants avoided using any of the language and ideas offered in the presentations and instead spoke of other things. The first speaker (a white woman) spoke of the need for safe spaces to enable teachers to be willing to try new things, and safe spaces for students to feel like they can succeed in science, but with no mention of race or African American students specifically. The next speakers followed her lead, speaking about students perceiving science and math courses as difficult, and the need for teachers to counter that belief. After allowing the discussion to flow in this direction for about 20 minutes, I suggested that science is perceived as hard by students of many social identities; however, as described in the presentations, African American students have an added challenge of largely negative images of them in society, images that separate them from inclusion in science. Introducing the racialized nature of identification with science yielded some reference among the participants to race; however, it was in the form of the idea that some teachers just "bond with some students and not with others." Despite repeated efforts to facilitate opportunities for the participants to apply the conceptual frames from the presentations to their experiences in schools, their reluctance continued. This avoidance of engaging with the offered language and ideas subverted the goal of considering the racialized nature of success in STEM learning.

Portraying Goodness

Another way that whiteness works to preserve and reproduce itself is through the desire or need to portray something good about whiteness (Martínez, 2017a). By doing so, it is possible for participants to see themselves as participating in a system that produces opportunity rather than oppression. These "positive" things attempt to show a more just and equitable side of whiteness and enable whiteness to be redefined so there is something good about it. This was evident in the working group conversations by a focus on sharing best practices.

During much of the time in the working group sessions, the participants gravitated toward sharing best practices, along with discussing the challenges of enacting such practices within scheduling and time constraints in schools. A long-time middle school teacher, who was white, took the lead in describing the many activities that had been used at her school to engage students, and often their families, in STEM. Ideas for projects were eagerly offered (most often by white educators) and gratefully accepted by others. Examples that were suggested included robotics and Lego competitions, bubble making, and building a golf course or a rain forest inside the school. While listening, one got the sense that the main thing holding back African American students in science lay in the lack of engaging, hands-on projects, along with time to plan and carry them out. It appeared as if sharing ideas and resources, and reconfiguring the school day or the school building, would solve the issue of lack of success of African American students in science. Sharing and collecting these kind of hands-on science projects and activities ensures their existence and proliferation in schools; however, it also reproduces the white story of schools as places where good things happen and of teachers as people trying to do good things.

Color Blindness

Espousing an ideology of color blindness is another way that the tool of whiteness is utilized. Doing so allows people to portray good intentions and assert that pedagogy is best when apolitical (Picower, 2009), while failing to address the systemic and institutional elements that lead to and perpetuate racism. This was seen in several instances in the working group discussions.

In the afternoon workshop session on the first day, I asked the participants to recall Dr. Mutegi's earlier talk, in which he suggested ways that teachers and schools could counter the negative images that assault African American students throughout our society. I focused their attention on his suggestions that schools could play a role in racial socialization and preparation for bias, and I asked the question of where these might fit into science teaching. Are there things related to racial socialization and preparation for bias that we could be doing as science educators, and how might we do them? Most of the suggestions offered by the participants, such as guest speakers to talk about careers, were color-blind and would be beneficial for all students, but did not address the additional burden of racialized storylines that falls on African American students.

When pressed on the role of teachers in positive racial socialization, there was no clear acceptance of that role. Several suggested it was the responsibility of family, church, and community. A white woman pointed to the need for parents to not contribute to a fear of math and science among their children, by claiming they were not good at math. This also is a color-blind suggestion that might benefit all students. In addition, this person actively rejected the racialized nature of students' feelings of competence in science or math, when she said, "I don't think it's a racial kind of thing. It's just a class and income thing that they just don't understand the importance of what it [science and math] is." Thus, color blindness was another way that whiteness supplied a means to deny, evade, subvert, and avoid the issues raised, such as the role of science education in racial socialization.

Dominant Narratives

Whiteness also functions to deny issues of race by covering up institutional racism through a reliance on particular dominant, mainstream stories that have become normalized and neutralized (Martínez, 2017a). In addition to color blindness, several other stories or narratives were evident in the working group discussions.

One of the key points in the readings that I provided for the participants and in my presentation in the opening session was the importance of teachers recognizing African American students' experiences and ways of thinking, talking, and being as resources on which science learning can be built. As explained earlier, I illustrated this with examples from my own research and from the work of others. I returned to this in the working group discussion, querying what they thought of this idea and if they had any examples of it in action. Most participants mentioned constraints that made it difficult to do this. Most commonly, the reluctance of teachers to welcome and build upon student contributions was attributed to a lack of confidence among teachers who do not feel that their content knowledge is adequate. In taking this stance, the participants were adopting a narrative of teachers being poorly prepared, a narrative about the teaching profession that is widespread and normalized. There was no apparent recognition by the participants that race or implicit bias might be a factor impacting the way the contributions of African American learners fail to be recognized and valued by a predominantly white teacher workforce. Therefore, the reliance on this majoritarian, mainstream story of underqualified science teachers resulted in covering up racism.

Similarly, dominant stories about African American students were employed by the participants. When one participant suggested having students read about or research scientists who look like them, another suggested that in her experience, "They did not have the basic skills of English. They did not have the basic skills of math," thus taking a deficit perspective that justifies continued remedial instruction for African American students.

Citing Exceptions

In response to Dr. Mutegi's powerful example of stereotype threat experienced by an African American university student, a white participant explained how she could not understand that reaction, because it was so different from how she would react. She went on to say that if she felt someone expected her not to do well, that would motivate her and make her work harder and excel. This response illustrates a lack of consideration of the racial difference between her and the student in Dr. Mutegi's example, thus, it is another example of a color-blind perspective. However, this example is even more complex in revealing how whiteness works. The participant's response also manifests a dominant narrative, that of rugged individualism. In addition, by introducing a counterexample or exception (in this case, the participant's own reaction), the speaker employs a common avoidance and rebuttal strategy. Sensoy and DiAngelo (2017) explain how citing exceptions or counterexamples allows people to avoid engaging with ideas in ways that deepen their understanding, and more importantly, such responses perpetuate the status quo and preserve the current, unjust system.

Silencing Communities of Color

Whiteness as a set of norms and perspectives embedded in an oppressive system is often taken up by all participants in that system—those who are white as well as those who are not, those who benefit greatly from the system and those who benefit less or not at all. Therefore, it was not surprising to see the tool of whiteness position white participants and their ideas in places of relative privilege and for that to be accepted and reified by others. In the working group discussions, when a question or prompt was posed by the facilitator, it was most often a white participant who spoke first. During conversations in which best practices were exchanged, it was usually white participants offering suggestions and black participants gratefully gathering them ("I'm going to steal that one. Thank you very much"). The unquestioned acceptance of these discursive positions led to a relative silencing of the participants of color.

Moving Beyond?

The reactions and responses described in the working group conversations are not unexpected, and to some extent, they have been described among educators elsewhere in the literature (e.g., King, 1991; Marx, 2004; Picower, 2009). As Marx found among the teachers included in her study, "the good intentions of the participants were consistently undermined by the whiteness and the racism" (p. 31) that remained unchallenged. The fundamental power of whiteness lies in its insidiousness, invisibility, pervasiveness, and perceived neutrality. Thus, it is difficult to question or challenge the perspectives of whiteness, as was evident in the working group discussions. Though conceptual tools had been provided to the participants, with the intention that those tools would facilitate the work of challenging whiteness and addressing why race matters in science and math education, those tools did not allow the participants to position themselves outside the system that whiteness protects. The tools offered by the presenters were no match for the powerful tools of whiteness. The propensity to portray goodness in the system and to call upon dominant narratives and counterexamples were powerful ways of denying, evading, subverting, and avoiding the issue of race in STEM education.

In dismantling whiteness, we need to do as Leonardo (2004) suggests and move beyond the idea that the system of racial oppression operates without the active participation of whites. The study of the working group conversations questions that underlying passivity and calls attention to how discursive practices actively protect and reproduce white supremacy. Through examining the kinds of deflections and avoidances illustrated in this chapter, we begin to see how talk and language, even among well-intentioned educators, can sustain a racially unjust educational system. However, by recognizing the many ways that whiteness does this evasive work and enabling educators to recognize their complicity in it, perhaps we can begin to see how to better challenge its protection of the status quo.

References

Boykin, A. W. (1986). The triple quandary and the schooling of Afro-American children. In U. Neisser (Ed.), *The school achievement of minority children: New perspectives* (pp. 57–92). Hillsdale, NJ: Erlbaum.

Emdin, C. (2010). *Urban science education for the hip-hop generation*. Rotterdam: Sense Publishers.

Frankenberg, R. (1993). *White women, race matters: The social construction of whiteness*. Minneapolis, MN: University of Minnesota Press.

Holland, D., Lachiotte, W., Skinner, D., & Cain, C. (1998). *Identity and agency in cultural worlds*. Cambridge, MA: Harvard University Press.

King, J. E. (1991). Dysconscious racism: Ideology, identity, and the miseducation of teachers. *The Journal of Negro Education, 60*(2), 133–146.

Lee, C. D. (2001). Is October brown Chinese? A cultural modeling activity system for underachieving students. *American Educational Research Journal, 38*(1), 97–141.

Leonardo, Z. (2004). The color of supremacy: Beyond the discourse of 'White Privilege'. *Educational Philosophy and Theory, 36*(2), 137–152.

Martínez, R. A. (2017a). 'Are you gonna show this to white people?': Chicana/o and Latina/o students' counter-narratives on race, place, and representation. *Race Ethnicity and Education, 20*(1), 101–116.

Martínez, T. (2017b, October 27). Academia, Love me back [Blog post]. Retrieved from https://vivatiffany.wordpress.com/2016/10/27/academia-love-me-back/

Marx, S. (2004). Regarding whiteness: Exploring and intervening in the effects of white racism in teacher education. *Equity & Excellence in Education, 37*(1), 31–43.

Moll, L. C., Amanti, C., Neff, D., & Gonzalez, N. (1992). Funds of knowledge for teaching: Using a qualitative approach to connect homes and classrooms. *Theory into Practice, 31*(2), 132–141.

Mutegi, J. W. (2013). "Life's first need is for us to be realistic" and other reasons for examining the sociocultural construction of race in the science performance of African American students. *Journal of Research in Science Teaching, 50*(1), 82–103.

Nasir, N. S., Snyder, C. R., Shah, N., & Ross, K. M. (2012). Racial storylines and implications for learning. *Human Development, 55*(5–6), 285–301.

Picower, B. (2009). The unexamined Whiteness of teaching: How White teachers maintain and enact dominant racial ideologies. *Race Ethnicity and Education, 12*(2), 197–215.

Putnam, R. T., & Borko, H. (2000). What do new views of knowledge and thinking have to say about research on teacher learning? *Educational Researcher, 29*(1), 4–15.

Sensoy, O., & DiAngelo, R. (2017). *Is everyone really equal?: An introduction to key concepts in social justice education*. New York, NY: Teachers College Press.

Steele, C. M., & Aronson, J. (1995). Stereotype threat and the intellectual test performance of African Americans. *Journal of Personality and Social Psychology, 69*, 797–811.

Van Den Bergh, L., Denessen, E., Hornstra, L., Voeten, M., & Holland, R. W. (2010). The implicit prejudiced attitudes of teachers: Relations to teacher expectations and the ethnic achievement gap. *American Educational Research Journal, 47*(2), 497–527.

Wang, M.-T., & Huguley, J. P. (2012). Parental racial socialization as a moderator of the effects of racial discrimination on educational success among African American adolescents. *Child Development, 83*(5), 1716–1731.

Warren, B., & Rosebery, A. S. (2011). Navigating interculturality: African American male students and the science classroom. *Journal of African American Males in Education, 2*(1), 98–115.

Toward a Transformative Framework for STEM Education: Achieving Equity Through a Holistic Approach

RONI ELLINGTON

Abstract

The persistent STEM achievement gap between black students and their white counterparts has been well documented in the literature. Although conversations about equity, inclusion and diversity have been in the forefront of our national discourse for the past 50 years, little progress has been made in closing this STEM achievement gap. One explanation for this lack of progress is that much of the literature in STEM education centered on Black students has been grounded in discourses of deficiency that positions black students as inferior, lacking social and cultural capital and having oppositional identities that impede their success in and persistence in STEM disciplines. Since the mid-1990s, however, there has been a sprinkling of research studies that have documented the voices of academically successful, historically marginalized students (Bergin & Cooks, 2000, 2002; Ellington, 2006, 2010; Fries-Britt, 2010; Moody, 1996). From these studies, we have begun to understand what promotes Black students' achievement and persistence and develop new school and classroom practices that can help deconstruct the "White male STEM myth" that is pervasive in STEM disciplines (Stinson, 2006). Research framed in discourse of achievement, although scant, has revealed many factors that foster success and persistence of Black students in STEM related disciplines and these factors have been used to inform a new framework for the STEM education of Black students. This chapter presents a Transformative Framework for STEM education that is grounded in a discourse of achievement that provides an assets based framework that can be used to inform school-based and classroom practices, and help address the persistent achievement gap documented in the literature. This chapter begins with a definition of the transformative framework for STEM education and its origins. Each component of the framework will be explored along with how these elements can be used in ways that can inform instruction and improve educational outcomes for Black students.

Introduction

As a graduate student in mathematics education during the late 1990s, I developed a passion for understanding how I could improve the educational outcomes of Black students in mathematics. This passion fueled my doctoral studies. Whenever I was given a project in my doctoral classes, I would use it as an opportunity to ground myself in the literature on Black students in mathematics and to learn how scholars were framing the discussion of the mathematics education of these students. I quickly discovered that much of the research on Black students in mathematics highlighted Black students' low performance, lack of achievement, and high attrition rates (Johnson, 1984; Oakes, 1990). The persistent narrative as it related to Black children in mathematics was that Black children were mathematically "broken" and needed to be fixed, and much of this "fixing" was to come from some form of curricular reform, new cognitive understanding or social intervention.

This unbalanced narrative of Black students' mathematics deficiency seemed odd to me, since I, as a Black student, had been successful in mathematics, had earned a master's degree in mathematics, and was pursuing a doctoral degree in mathematics education. Since I did not feel as if I were "special" in any way, there had to be other successful Black mathematics students who could provide an alternative narrative. If the narrative of Black students in mathematics was as bleak as was being described, why had researchers in mathematics education not solicited the views and experiences of successful Black mathematics students to reframe the deficit-based discussion to an assets-based one that could be used to promote success for other Black students? I knew intuitively that studying failure was insufficient for understanding success, and this insight sparked my interest in investigating the experiences of Black students who had excelled in mathematics, mainly as mathematics majors. This investigation provided the foundation for the framework being presented in this chapter.

In my initial research, I had sought to understand the personal, educational, social, and cultural factors that high achieving Black students perceived to be instrumental to their success (Ellington, 2006). The findings of this study uncovered many factors to which they attributed their success and persistence in mathematics. These factors included: resiliency, positive mathematics identities, the presence of positive role models, parental support and advocacy, spiritual and social consciousness, and peer support. In that early research, I offered recommendations for how such factors could be understood and integrated into our current social, cultural, and educational practices. My original research was focused explicitly on mathematics majors and the implications for mathematics education. However, since much of the conversations in the literature began focusing more broadly on STEM education, I started looking for ways to apply my work to this emerging discourse.

It was during this time that I became involved in several projects that focused on interdisciplinary science and mathematics learning, and I began to review the broader STEM education literature. I discovered that this deficit-based narrative of Black students was not confined to mathematics education, but was the dominant narrative in the STEM education literature as well. The STEM literature documented the fact that Black students have persistently underperformed White students in STEM-related disciplines (Berry, 2003; Lubienski, 2002) and fail to persist in STEM disciplines at the rates of their White counterparts (Jencks & Phillips, 2011; Vanneman, Hamilton, Anderson, & Rahman, 2009). Ironically, although conversations about equity, inclusion, and diversity have been at the forefront of the STEM national discourse for about 30 years, little progress has been made in closing this STEM achievement gap. In fact, some scholars have argued that things have become worse (Geertz, 2003).

Origins of a New Framework for STEM Education

I discovered that in STEM education just as in mathematics education, many of the programs and practices advocated by proponents of equity in STEM were framed using a deficit model, a perspective of Black children as inferior to White students due to personal, institutional, or cultural deficiencies. As Stinson (2006) asserts, "The discourse of deficiency focuses on the perceived deficient cultural, schooling, and life experiences in general, of Black children. Educators who participate in this discourse often claim that the low performance of many Black students exists because Black children experience higher rates of poverty, live in high-crime communities, and hail from unstable single parent homes with minimal parental involvement" (p. 483). Also, this deficit-based discourse argues that Black students are "traumatized" and suffer from the negative effects of slavery, segregation, racism, and discrimination that impede their ability to be successful in school (Smith, Allen, & Danley, 2007). This deficit-based view of Black children and the pedagogical practices grounded in this perspective have led to school-based programs and interventions, designed to "close the achievement gap," which have not worked, have not addressed the needs of Black children nor do they draw on the brilliance of these children (Leonard & Martin, 2013). Further, these programs and practices have not been comprehensive enough to effect real change in the achievement and persistence of Black children in STEM disciplines (Gutiérrez, 2008; Ladson-Billings, 2006). In fact, many scholars have argued that these efforts have in fact undermined our diversity and inclusion efforts (Martin, 2010) and have served to further marginalize Black students in STEM-related disciplines (Martin, 2016). Hence, there is need to reframe equity in STEM education for Black students using a discourse that affirms the individual and collective assets that Black students bring to STEM-related learning experiences.

A counter-discourse, although still emerging, is the *discourse of achievement*. To date, this is still the least researched and theorized discourse in STEM education (Stinson, 2006). Since the mid-1990s, however, there has been a sprinkling of research studies that have documented the voices of academically successful, historically marginalized students (Bergin & Cooks, 2000, 2002; Ellington, 2006; Ellington & Frederick, 2010; Fries-Britt, Younger, & Hall, 2010; Moody, 2001). Research framed in *a discourse of achievement*, although scant, has revealed many of the factors that foster success and persistence of Black students in STEM-related disciplines (Ellington, 2006; Ellington & Frederick, 2010; Moody, 2001, 2004). It is this discourse that has shaped the framework that I am proposing in this chapter. In the remainder of this chapter, I define a transformative framework for STEM education and discuss each component of the framework, its research-based underpinnings, and several considerations for pedagogical practice that could inform instruction. STEM education as currently practiced has been ineffective in decreasing the achievement gap and fostering high achievement and persistence in STEM disciplines for Black students (Huberman, 1991); hence, it is important that educators incorporate instructional practices that address Black students' needs. To do this, these practices must be grounded in a holistic approach that draws on the strengths of Blacks students and their communities, fosters maximal interest, sustained engagement, and exceptional achievement for these students.

Current approaches to reforms in STEM education appear to focus narrowly on curriculum reform rather than providing a holistic approach that could result in sustainable changes. Also, these approaches offer a mere "tweaking" of a STEM educational system that has consistently failed Black students. For example, many efforts to promote STEM interest and achievement for underserved students often add enrichment activities to the curriculum such as robotics, coding, computer programing, engineering, and interdisciplinary learning experiences (Lee, 2015; Mihyeon, Cross, & Cross, 2017). What these approaches neglect to do is address the context in which Black students exist, or address the complexity of STEM education, the societal messages being sent to discourage Black students from engaging, achieving, and persisting in STEM disciplines, and how Black students internalize these messages. What is most egregious is that when these programs and practices "do not work" or the results are not sustainable over time, Black students are further marginalized because the deficit thinking often gets solidified through the failure of these efforts (Martin, 2012). Hence, it is imperative that we rethink not only our practices but our perspectives, beliefs, and messages that underlie our STEM education system. Programs grounded in a holistic approach will do more than change the trajectory of Black students' STEM achievement but transform who we are, how we approach our work, and the long-term educational outcomes for these students.

It is with this transformational view in mind that I have developed a Transformational Framework for STEM education, which was created from my initial research on the experiences of high achievers, an analysis of literature grounded in an asset view of STEM education, and discussions with national leaders in STEM education. **A Transformative Framework for STEM Education** has four interactive components: 1) student identity, agency, sense of belonging, and resiliency; 2) utilization of, and contribution to, community social and cultural capital; 3) transformative school practices including curricular, instructional, and assessment reform; and 4) teacher professional development focused on teacher reflection, social consciousness, and empowerment. All four components of the framework must be deliberately and intentionally integrated into any STEM education program to ensure that its goal of equity and inclusion are realized. In the next sections, I discuss each component of the framework and how it can be integrated into STEM programs and practices that support Black learners to excel and persist in STEM.

A Transformative Framework for STEM Education

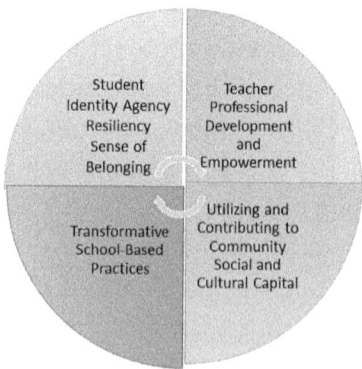

Figure 2.1: A Transformative Framework for STEM Education (source is author's own).

Student Identity, Agency, Sense of Belonging, and Resiliency

Students are central to the teaching and learning of STEM, hence understanding who students are and what experiences support their successful navigation of the STEM pipeline is essential to any useful framework. Fundamental to understanding the central role Black students play in their own success in STEM is a clear understanding of identity, agency, sense of belonging, and resiliency, and how these impact student learning, achievement, and persistence in STEM.

Several scholars who study Black students in STEM have studied the concept of *identity* to gain insight into its role in positive student outcomes (Cook, 2014; Kane, 2012; Martin, 2012; McGee, 2015; Wilson, 2016b). Although there are many definitions and conceptions of identity (Abrams & Hogg, 1988; Jenkins, 1996; Oyserman, 2007; Oyserman, Elmore, & Smith, 2012), it is broadly conceptualized as one's answer to the question "who am I?" or the ways in which individuals and groups define themselves, are defined by others, and relate to others (Abrams, 1994; Abrams & Hogg, 1988; Deng, 1995; Oyserman, 2007, 2008). As stated by Oyserman *et al.* (2012), "One's sense of self (his or her identity) influences what he or she is motivated to do, how one thinks and make sense of oneself and others, the actions one takes, and one's feelings and ability to control or regulate oneself" (p 74). Black students, as do all students, possess different types of identities (e.g., social, racial, gender, religious, academic), and they negotiate these different identities within different contexts. Further, scholars argue that these identities are fluid and not fixed (Gergen, 1991). Researchers suggest that learner's identities are enacted and shaped by their participation in socially situated practices such as STEM education (McGee, 2015). Hence, central to the promotion of Black students' persistence and success in STEM is an understanding of the types of socially situated STEM practices (community, school based, classroom) that foster identities that will promote their success and persistence in STEM.

Highly connected to identity and frequently used synonymously with identity is the term *self-concept*. Self-concept can be defined as "cognitive structures (content, attitudes, or evaluative judgments) that are used to make sense of the world, focus attention on one's goals, and protect one's sense of basic worth" (Oyserman & Markus, 1998). As Armstrong-West and H. de la Teja (1988) argue:

> An individual's idea of who she or he is—the self-concept—contributes significantly to how the individual responds to society's institutions. The extent to which a person's self-concept is confirmed or rejected by others is critical to the person's development and social and academic integration. (p. 36)

Positive self-concept and STEM identity have been consistently linked to positive STEM-related outcomes for Black students (Brown, Mangram, Sun, Cross, & Raab, 2017; Varelas, Martin, & Kane, 2012), specifically their willingness to engage and persist in STEM. There is overwhelming support in the literature for considering student identity in the development of STEM programs and practices targeted explicitly for Black students (Brown *et al.*, 2017; Kane, 2012; McGee, 2015; Vareles, Martin, & Kane, 2012). Specifically, Black students in the STEM pipeline are continuously developing, negotiating, defining, and redefining their various identities in ways that answer the question "Who am I as a STEM learner?" and "How does the pursuit of STEM affirm or deny my sense of self?" Hence, any framework that is being used to shape STEM education for Black students must include a clear explication of how the programs and practices employed affirm Black children's sense of self and promote positive social, racial, and STEM-related academic identities.

In addition to a STEM framework that promotes positive identity and self-concept, a comprehensive STEM framework must consider how to cultivate and expand Black students' agency. Personal agency can be defined as one's capacity to originate and direct his or her actions for a given purpose (Zimmerman & Cleary, 2006). As Black students develop positive identities and self-concepts that affirm them as competent STEM learners, they must learn to think for themselves, powerfully navigate the STEM pipeline, and act in ways that can emancipate them from the inevitable barriers that they will experience as Black STEM students (Ellington, 2006). These are key aspects of one's agency. In his pivotal study of successful Black students in mathematics, Martin comments:

Although I refer to a range of dispositional factors and strategies that contributed to their success as important, an important and often neglected component [in these Black students' success] was a personal agency. A particularly noteworthy finding [of the current study] was the degree to which successful [black] students recognized and responded productively to their surroundings.

Other scholars have also found that Black students' agency was critical to their success in STEM-related disciplines (Long & Henderson, 2017) and is a critical personal characteristic of Black students who excel in STEM courses and careers (Ellington, 2016; Ellington & Frederick, 2010). Hence, it is critical that equity-focused STEM programs include practices that help develop students' agency which will help them powerfully navigate an often unwelcoming STEM environment (Long & Henderson, 2107).

Some scholars suggest that one's sense of agency is impacted by one's sense of belonging in the STEM community (Johnson, 2012). A sense of belonging refers to a student's feeling that they fit in with the people, materials, and activities within an environment (Cheryan, Plaut, Davies, & Steele, 2009), in this case, the STEM environment. The extent to which Black students perceive a disconnect between the academic environment and their own identity, the less likely they are to feel that they belong there, and hence are less likely to achieve and persist in the discipline (Cheryan *et al.*, 2009; Green, Emery, Sanders, & Anderman, 2016; Johnson, 2012). Belonging is thought to be fundamental to human motivation, and a lack of belonging can lead to adverse effects on academic motivation, personal agency, and/or sense of wellbeing (Baumeister & Leary, 1995). A sense of belonging has been shown to be a strong predictor of Black students' interest, motivation, and persistence in STEM disciplines (Harper, 2012; Strayhorn, 2015). Hence, any program or practice designed to transform Black students into effective STEM learners must address how students' personal agency and sense of belonging are being cultivated in the program and must be done intentionally, thoughtfully, and deliberately.

Closely related to personal agency is the concept of *resiliency*, which has been linked to Black students' ability to remain steadfast in their academic pursuits in the face of a myriad of challenges they inevitably face (Andrews, 2012; Getz, 2000; Goodwin, 2002; Long & Henderson, 2017). Resilience has been defined as "a set of inner resources, social competencies, and cultural strategies that permit individuals to not only survive, recover, or thrive after stressful events but also to draw from the experience to enhance subsequent functioning" (Stanton-Salazar & Spina, 2000). Much of the literature that describes the experiences of successful Black students in STEM has identified resilience as a key element in their success. Scholars have found that successful Black students develop the inner resources to not only survive but thrive in STEM-related contexts that are often uncomfortable and even hostile toward them. Some scholars have used terms such as "grit" (Duckworth, Peterson, Matthews, & Kelly, 2007) "perseverance," and "personal determination" to describe the ability of Black students to persist and overcome the obstacles they inevitably face in the STEM pipeline (Andrews, 2012; Sandoval-Lucero, Maes, & Klingsmith, 2014).

Although much has been said about the importance of resiliency, many reform-based practices used to promote equity in STEM do not deliberately cultivate Black students' resiliency nor do they make students aware of how they can draw on their resiliency to overcome the obstacles they face as STEM learners. Any viable framework must include an understanding and application of resiliency, particularly how instructional leaders can deliberately cultivate it in Black students. Unfortunately, many scholars describe resiliency as a trait that students either possess or fail to possess or as something that is the students' responsibility to cultivate. By describing resilience in this way, the development of this "trait" has been left for students to cultivate on their own without a deliberate effort to support students to develop resiliency through well-designed instructional experiences. Further, students who do not display characteristics of resiliency or "grit" are often viewed as deficient, and this has justified our unwillingness to develop instructional practices that cultivate perseverance, diligence, and resiliency.

To summarize, understanding Black students' identity, personal agency, sense of belonging, and resiliency is central to developing a framework that will nurture their success in STEM. Any approach to redefining STEM education that does not consider how the program or practice integrates these concepts into the teaching and learning of STEM will fall short of meeting its ideal outcomes. The fundamental message that our STEM practices should reiterate to Black students is,

> "We are competent and confident lifelong STEM learners, doers and community members. We contribute to our communities through our STEM learning."

Some considerations that might inform the development of pedagogical practices that can be used to foster positive identities, personal agency, community, and resiliency for Black students include:

- Developing STEM education that encourages learning as participation not merely as an acquisition of knowledge
- Using practices that promote academic and social integration for Black students
- Integrating opportunities for Black students to draw on their agency and resilience in their STEM learning with the assumption that agency and resiliency is inherent in all children.
- Incorporating self-efficacy and self-concept development strategies and a sense of belonging/community
- Creating STEM education that liberates and empowers, not just educates
- Incorporating opportunities for peer support and student-directed instruction
- Intentionally fostering "grit" and resilience through well-designed instructional activities

Utilizing and Contributing to Community Social and Cultural Capital

Another critical aspect of a transformative framework for STEM education is creating STEM education that utilizes and contributes to the social and cultural capital of students' communities. For the past five decades, the social and cultural capital literature has been used to understand students' ability to succeed and effectively navigate schooling (Thirutnurthy, Kirylo, & Ciabattari, 2010; Franklin, 2004), specifically STEM education (Cegile & Settlage, 2014; McAllister, 2011). Unfortunately, this discourse has focused on the "lack" of social and cultural capital that Blacks and other minorities have, given the "negative plight" of these communities. The social and cultural capital discourse is generally framed in ways that advance a deficit orientation of Black students and their communities. However, there is a growing body of scholarship in STEM that is focusing on an asset-based view of Black students' social and community resources. This emerging body of research highlights ways to empower Black students in STEM by understanding the parental, community, and personal resources that they bring to their STEM learning and drawing on these resources to improve educational outcomes for Black students (Ellington, 2066; Ellington & Frederick, 2010; Harper, 2010; Stinson, 2006).

To fully understand the importance of a STEM education that utilizes and enhances Black students' community assets, it is important to understand the traditional conceptions of cultural and social capital and how these conceptions need to be reframed to promote success and persistence in STEM for Black children. Although there are many conceptualizations of social and cultural capital (Bourdieu, 1996; Coleman, 1988; 2000 Franklin, 2004; Lin, 2001), researchers generally contend that social and cultural capital can be invested and mobilized by a group or individual to yield positive outcomes in society such as wealth, power, or reputation (Lin, 2001). Specifically, cultural capital refers to systems of attributes, dispositions, language skills, and cultural beliefs, values, and knowledge inherited from one's parents (Bourdieu, 1986). Cultural capital can exist in three states: *embodied* (dispositions of mind and body), *objectified* (cultural goods), and *institutionalized* (educational qualifications) (Bourdieu, 1996). Specific forms of cultural capital are valued more than others, and each person brings a different set of dispositions to any field of interaction such as schooling.

Related to cultural capital is the concept of social capital. Coleman's (1988) interpretation of social capital is the most frequently cited in the educational literature. For Coleman, social capital is defined by its function. Coleman asserts that it (social capital) is not a single entity, but a variety of different entities which have two characteristics in common:

they all consist of some aspect of social structure, and they facilitate specific actions of individuals who are within the structure' (Coleman, 1990, p. 302). Coleman proposes that social capital is intangible and has three forms: (a) level of trust, as evidenced by obligations and expectations; (b) information channels; and (c) norms and sanctions that promote the common good over self-interest. What is important to note about both cultural and social capital is that they can be accessed and used by a group or individual to promote positive educational outcomes.

Historically, the more the values, dispositions, beliefs, and social networks of an individual or group reflect and draw upon those of the dominant culture, the more social and cultural capital that an individual or group possesses. Some scholars refer to cultures that reflect this congruence as "elite status cultures" (DiMaggio, 1982). Literature suggests that educational systems reward students who participate in these "elite status cultures," which primarily consist of White, middle class, male, able-bodied, and native English speakers and devalue students who come from cultures that don't fit these normative cultural perspectives and ideologies. As a result, students from these elite status cultures are given more attention and special assistance and are perceived as more intelligent or gifted than students who lack cultural capital (DiMaggio, 1982). Hence, participation in these "elite status cultures" represents a kind of cultural capital, while not being in these cultures can cause one to be marginalized and even ignored throughout every phase of the educational pipeline.

Unfortunately, the communities in which Black students reside have not been viewed as "elite status cultures" given the mismatch between their values, cultural norms, practices, and social networks and those of the educational system, specifically the STEM culture (Ogbu, 1978, 2003). In fact, Black students' communities have been labeled deficient, therefore research focusing on them highlights the perceived deficient social, cultural, schooling, and life experiences of Black students (Stinson, 2006). Scholars who engage in this discourse often claim that the "lower" academic achievement of many Black students exists because Black children hail from communities that lack the social and cultural resources needed to be successful (Ogbu, 1978). For Black students to engage and succeed in STEM, we must reframe this deficit view of Black students' social and cultural capital to one that views Black students' communities as rich sources of capital that can be utilized in their STEM learning. Research and practices grounded in this assets-based view would start from a premise that Black communities are rich in academic, social, and knowledge resources that can be used in STEM education.

For STEM education to be a viable and meaningful pursuit for Black learners, it must draw on the lived realities of Black students and the various forms of capital they bring to STEM learning. To illustrate this, I return to Bourdieu's (1986) three states of cultural capital: embodied (dispositions of mind and body), objectified (cultural goods), and institutionalized (educational qualifications) and the three forms of social capital proposed by Coleman (1990), which include (a) social levels of trust, (b) information channels, and (c) norms and sanctions that promote the common good over self-interest. In the remainder of this section, I argue that Black students bring these states and forms of social and cultural capital to STEM learning, and it is essential that programs and practices designed to promote their success and persistence in STEM be aware of and incorporate these into their pedagogical practices. Also, well-designed STEM learning experiences that enhance these forms of capital can empower students to give back to their communities in ways that can help these communities thrive. Hence, Black students' STEM education will be mutually beneficial to themselves and their communities, which is essential for transformative outcomes.

Researchers have noted that Black students have a high sense of social responsibility, value giving back to their communities, and a broad sense of spirituality (Ellington, 2016; Ellington & Frederick, 2010). These factors influence social networks, social norms, and levels of trust that can be utilized for STEM learning. Hence, it is crucial that we provide ways for Black STEM students to understand how their STEM learning experiences can contribute to their communities, promote social activism, and connect them to a higher purpose (Walker & Dixon, 2002). STEM educators can build on these dispositions and the social norms that promote the common good and not just individualistic achievement. Such an approach is likely to ignite STEM interest and fuel a passion for STEM learning.

In addition to favorable dispositions and viable social networks, Black students' communities have cultural goods, institutions, and social systems that can be used to foster their STEM success. Specifically, Black parents, community mentors, and business leaders can be utilized in meaningful and mutually beneficial ways and can serve as viable partners in our STEM programs. In a deficit-based model, the cultural goods and social networks of Black communities have been minimized. In fact, the primary message underlying these efforts has been that we need to "give something to Black students and their communities that is inherently lacking," and this message undermines real and mutually beneficial partnership-building with these communities.

For example, a common approach to providing enrichment programs has been the development of school-based, after school, or out-of-school learning experiences in such areas as coding or robotics to enhance STEM learning. Many of these efforts are deployed with little or no input from these communities (Vasir, 2104). The assumption is that these communities do not have assets in these areas and that they require an outside entity to provide them with these interventions. As discussed by Nasir and Vakil (2017), "[STEM learning environments] fail to connect the learning of coding, design or STEM more generally to broader social and racial justice issues. Code for what? For whom? For what purpose? Whose problems are being coded for? Which questions are being asked?" Further, it is also assumed that the communities desire these programs and that they reflect a need expressed by the community (Alvarado & Muniz, 2015). In many instances, this is not the case. It is not to say that these programs are not or cannot be viable. The point is that many in-school and out-of-school STEM programs are not generally developed and implemented in partnership with the community and in ways that can maximize the cultural and social capital already embedded in these communities. As a result, these initiatives can fall short of their potential to transform learning outcomes for Black students. It must be noted that Black communities have a wealth of churches, businesses, organizations, social leaders, parents, and advocates that could be embraced and engaged in all phases of STEM education.

It is also essential that the STEM education of Black students empower them to be agents of change and advocates in their communities. Specifically, scholars in the social justice tradition of STEM education (Apple, Au, & Gandin, 2009; Gutstein, 2006, 2007; Nieto & Bode, 2008) have advocated for a STEM education that addresses the real issues facing Black students and their communities. In the results from a meta-analysis of the effects of out-of-school time (OST), Young, Ortiz, and Young (2017) found that STEM interest is not sufficiently developed in out-of-school settings that lack a social focus. They recommended that researchers and policy makers design OST programs that "develop leadership skills and interactions between the OST settings, student homes, communities, and the business sector, which can foster a social and emotional connection that students can draw from as motivation to pursue and complete degrees in STEM fields" (p. 67). They go further to suggest that "students of color do not recognize STEM fields as platforms to reach their altruistic goal of helping others, which contributes to their decision not to choose a STEM-related career. Hence, there must be a concerted effort of educators to have STEM education be used to enhance their lives, their communities and solve real issues facing their communities" (Young et al., 2017, p. 69). Further, there is a need to address how schooling and educational structures contribute to creating an unjust society and how teachers and educational leaders can work toward social change by addressing these inequities (Apple et al., 2009; Nieto & Bode, 2008).

As scholars have suggested, STEM education cannot be viewed as a set of isolated facts and disjointed skills that have little or no relevance to the lived realities of students (Gutstein, 2006, 2007). One way to address this lack of coherence and relevance is to build STEM programs around the social and cultural assets that the communities possess and develop learning experiences that engage the community, draw on its social values, norms, and resources, and reflect the social justice aims of STEM learning. The message that must be sent to Black students is that

We recognize that education occurs in historical, cultural, and social contexts, and we draw on the rich resources of the community to enhance the teaching and learning in STEM disciplines. We build mutually beneficial partnerships with communities that draw on the social and cultural capital of Black students and their communities to foster STEM success and persistence for all students and give back to these communities in meaningful and sustainable ways.

Considerations for pedagogical practices that can be implemented to help send this message include:

- Engaging in authentic and sustainable partnership-building activities
- Providing mentoring and academic support programs for Black students
- Creating learning experiences connected to community concerns and issues
- Using STEM education to empower and liberate Black students and communities
- Developing partnerships with businesses
- Having entrepreneurial STEM activities and programs

Transformative School Leadership, Curriculum, Instruction, and Assessment Practices

Effective school leadership is essential to all students' success and persistence in schooling but more so for Black students in STEM. School-based practices that have been shown to motivate, inspire, and educate Black students to engage and succeed in STEM and help them understand the importance and usefulness of STEM to their lives and their communities require intentional and informed support from school leaders. Essential to a framework that fosters success and persistence for Black students in STEM education is the development and implementation of school-based leadership, that is grounded in culturally responsive, social justice, critical theory, and African-centered pedagogy (Nasir & Vakil, 2017). This includes support for reformed curricular, instructional, and assessment practices.

Scholars have documented the importance of school leadership in creating effective schools and promoting positive student outcomes (Ford, 2018; Ward et al., 2015). School leadership has been shown to be a critical factor for creating and sustaining functional schools (Robinson, Lloyd, & Rowe, 2008) and promoting success for Black students (Shields, 2012; Wilson, Douglas, & Nganga, 2013). Also, empirical evidence indicates that leadership has a positive effect on student learning as well as on STEM persistence (Gutstein, 2006). For example, Ford (2018) conducted a study of four exemplary Inclusive STEM-focused high schools (ISHSs), identified by experts in STEM education as highly successful in preparing students who are historically underrepresented in STEM. He identified various internal and external leadership factors influencing school leadership and examined an existing data set that focused on the leadership contributions of ISHS leaders and their broader community. One of the significant findings of the study was that school leadership was essential to the success of these schools. The leadership employed in these schools reflected a concept of school leadership that included within and outside of school leadership, autonomy and innovation, responsiveness to their schools' needs, and an ongoing investment in increasing their schools' capacities, which included investing in teacher professionalization, providing pathways for school leadership, collaborating with business and industry, and identifying the best student supports.

As this study and other literature suggest, school leaders must consider how their leadership promotes Black students' success in STEM. Encouraging innovation, adapting school-based programs that have been shown to promote viable STEM learning including interdisciplinary learning, STEM academies, social justice-focused STEM experiences, and integrated STEM instruction are elements of leadership that have been shown to have a positive impact on school culture and student success. Most importantly, school administrators must advocate for teachers and students, partner with communities, and allocate resources for professional development for teachers and innovative in-school and enrichment programs for STEM learning for their students (Bryk, Sebring, Allensworth, Luppescu, & Easton, 2010; Ford, 2018).

Research on successful STEM schools for Black students asserts that school leaders must integrate a focus on STEM with a transformative approach to leadership that focuses on a community of caring (Bass, 2012; Milner, 2013; Thompson, 1998; Wilson, 2013, 2016b), teacher and student empowerment, and attention to issues that impact Black students, including power, privilege, positionality, race, and racism (Shields, 2015). As Wilson (2016) asserts: "Transformative educational leadership is a political process that requires educators to understand that schools should be sites of resistance and encourage social justice where they work to redress inequities" (p. 558). Transformational leaders guide staff and students in "deconstructing and reconstructing knowledge that values diverse and critical perspectives; develop inclusive and affirming learning communities for marginalized students; and, serve the public good" (Wilson, 2018, p. 558). Tillman (2004), in her comprehensive review of research on African American principals, emphasizes how "transformational" African American principals commonly demonstrated "interpersonal caring" or an "ethic of care" along with political resistance against injustice (p. 124). Moreover, transformational leaders advocate for teachers and students in the face of injustice, given their commitment to caring for their students and promoting equitable outcomes for all students.

Guided by useful school-based STEM models and a focus on transformational leadership, school leaders can foster the kind of learning environments that encourage positive learning outcomes for Black STEM students. Leaders who understand and integrate STEM learning with caring teachers, student empowerment, and social justice are critical to teachers' ability to adopt these transformative practices in their classrooms. Research shows that Black students succeed and persist in STEM disciplines when they engage in meaningful, useful, and empowering STEM learning experiences that are grounded in culturally responsive, antiracist, and African-focused curricula, instructional, and assessment practices.

Culturally Responsive Teaching (CRT), Culturally Responsive Pedagogy (CRP), or Culturally Relevant Pedagogy can be defined as using cultural characteristics, experiences, and perspectives of diverse students as conduits for teaching them more effectively (Gay,

2002). Although there are many frameworks for CRT (Gay, 2002; Ladson-Billings, 1994, 1995a, 2001; Vilegas & Lucas, 2002), essential elements of CRT include the following: active learning is encouraged, curricula materials and activities are connected to the lives of students, and classroom environments are inclusive and interactive (Montgomery, 2001). One of the key aims of CRT is to "empower students intellectually, socially, emotionally and politically by using cultural referents to impart knowledge, skills and attitudes" (Ladson-Billings, 1994, pp. 17–18). This empowerment is achieved through the integration of home and community experiences with school, curriculum, and classroom. Looking specifically at African American students and their teachers, Ladson-Billings (1995) found that CRT relies on three foundational premises: "(a) students must experience academic success, (b) students must develop and maintain cultural competence, and (c) students must develop a critical consciousness" (p. 2). Hence, CRP (Ladson-Billings, 1994) offers practical strategies for the academic success of Black students that have been applied to STEM education by various scholars (Dimick, 2012; Hubert, 2013; Johnson, 2011; Laughter & Adams, 2012; Leonard, 2017; Rodriguez, Bustamante Jones, Peng, & Park, 2004; Young, Young, & Hamilton, 2013). These studies have shown positive learning outcomes related to STEM learning for Black students exposed to CRT.

Although scholars in STEM education continue to advocate for integrating CRP in STEM, some scholars suggest that this pedagogy does not go far enough to dismantle the structural inequities that impact Black students and doesn't adequately address the issues of race and racism that underlie Black students' lived realities, their education, and their progression in the STEM pipeline (Asante, 1991; Murrell, 2012; Shockley, 2007; Shockley & Cleveland, 2011). A growing number of scholars are therefore advocating for pedagogies that interrogate issues of power, positionality, privilege, and White supremacy and their impact on Black students. Some scholars argue that there is a need to go beyond CRT to pedagogies that make Black students' history, culture, and experiences central to their learning and move beyond culturally responsive to more African-centered, antiracist, and race central pedagogies (Murrell, 2012; Shockley, 2007). As Shockley (2007) argues, "The Afrocentric approach in education involves working with Black students to master the academic disciplines from a perspective that grounds them in an African reality. That means children are taught about events, places, people and things, with essential reference to and in the critical context of the historical trajectory of people of African descent. Proponents of Afrocentric education wish to impress upon African American parents the desperate need for an educational experience that includes proper cultural grounding" (p. 55). Research integrating the tenets of these pedagogies with STEM education is scant (Lord-Walker, 2015; Thompson & Davis, 2013). However, there is evidence to suggest that engaging Black students in STEM instruction that centralizes race, racism, and the Black experience nationally and globally is transformative for Black students and their communities (Lord-Walker, 2015) and should be included in any STEM educational framework that will address the needs of Black students.

STEM educational experiences at all levels should reflect a deliberate concern for how students' lived realities can inform STEM knowledge. The goals of STEM education must be expanded from the conventional notion of preparing Black students for work to one that empowers them to make an impact on themselves and liberate their communities (Jett, 2009). This reframing of STEM education will require educators and instructional leaders to provide more than "lip service" to the relevance of STEM or merely tout an individualistic benefit for STEM learning. This reframing allows students to see how their STEM learning can be used to critique the status quo and contribute in ways that are aligned with their passions and future positions in their communities. This reframing of STEM education will urge STEM educators and advocates to adopt models of STEM education that are interdisciplinary, integrative, and empowering for Black students.

The message that must be sent to Black students should be

> STEM Education is Critical, Vital and Relevant to each of us, our communities, and the world and can be used to interrogate inequities and offer solutions to real-world issues that impact our communities. We use STEM knowl-

edge to solve real-world problems, understand the world, contribute to the larger society and liberate ourselves and our communities.

Some considerations that should inform transformative school leadership include:

- Developing Constructivist, Humanist, and Critical Paradigms of Teaching and Learning
- Creating a school "Climate of Care"
- Developing and Adopting Transformational School Leadership Models for STEM schools
- Focusing on Culturally Relevant STEM Curriculum and Pedagogy
- Building viable partnerships with the Black Community that inform STEM curriculum and instruction
- Adopting Interdisciplinary and Transdisciplinary Teaching and Learning Models in STEM-focused schools
- Integrating Problem-Based Learning experiences throughout the curriculum
- Critiquing Dominant Narratives in STEM education
- Empowering Teachers and Students to implement innovative STEM curriculum, instruction, and assessment practices

Teachers' Professional Development: Competent STEM Educators as Empowered Agents of Transformation

For STEM education to be equitable for Black students, teachers must be able to deliver instruction that is innovative and engaging, student-focused, grounded in the social and cultural capital of their communities, and reflect transformative practices in STEM education. For teachers to be able to do this, they must be competent in STEM pedagogical practices as well as practices that promote equity, empowerment, and liberation for Black students. In order to achieve this, teachers must engage in professional development that reflects principles of high-quality

STEM professional development (Asghar, Ellington, Rice, Johnson, & Prime, 2012; DeSimone, 2009; Loucks-Horsley, Hewson, Love, & Stiles, 1998; Wojnowski & Pea, 2014), fosters critical reflection (Howard, 2003), develops a critical consciousness and empowerment (Gay & Kirkland, 2003; Ellington, 2016), and forwards a social justice and antiracist agenda (Joseph, Haynes, & Cobb, 2016). STEM educators must be provided professional development that not only makes them effective at teaching STEM content but also develops them into critical and reflective practitioners who see their role as STEM educators as more than content specialists but as agents of transformation and advocates for Black children.

The literature on teacher professional development, specifically the professional development of STEM teachers, has identified the factors that characterize effectiveness. High-quality professional development in STEM must model inquiry approaches and emphasize the importance of subject-matter content knowledge and deep conceptual understanding (Asghar *et al.*, 2012). Models of professional development that produce highly effective STEM teachers should foster active learning on the part of teachers and students, incorporate reform-based content and pedagogy, encourage teacher collaboration, and must be ongoing and integrated into their work (Wojnowski & Pea, 2014). Categories that encompass the professional development of mathematics and science teachers include immersion (teachers actively involved in doing mathematics and science); curriculum (engaging teachers with learning materials that they will use with their students); examining practice (PD that is job-embedded); collaborative work (professional networks and learning communities), and vehicles mechanisms (PD structured primarily through workshops and institutes) (Loucks-Horsley, Hewson, Love, & Stiles, 1998). Although many models and approaches to STEM professional development have been developed over the years, the primary goal of professional development has remained constant: to change teacher practice and improve student performance (Wajnowski & Pea, 2014). By incorporating these principles into STEM teacher professional development, research shows that teachers are more effective at reaching these goals (Asghar *et al.*, 2012; Grigg, Kelly, Gamoran, & Borman, 2013; Laura, McMeeking, Orsi, & Cobb, 2012).

In current reform efforts, the goals of professional development have expanded to include the empowerment of teachers to change their practice in ways that promote equity, inclusion, and high achievement for all students, specifically those who have been historically marginalized. As Nieto (2006) suggests, although qualities of effective teachers such as content knowledge and pedagogical knowledge are essential for teachers to be successful in the classroom, other characteristics such as a passion for social justice and the courage to challenge "the status quo" are essential given the sociopolitical contexts of schools and the need to address the inequities that exist. Hence, there is a need for an expanded view of STEM professional development. This need moves beyond developing teachers' content and pedagogical knowledge to supporting teachers to be reflective practitioners who embrace student cultures and identify their own cultural biases, so that their insecurities will not hinder their ability to give every student [particularly Black students] expanded opportunities to learn (Young *et al.*, 2013).

It is critical that STEM teachers who teach Black students be reflective practitioners who are willing to not only reflect on their teaching practices but are also able to reflect on who they are, their beliefs and perspectives, and the various biases they bring into the classroom (Gay & Kirkland, 2003). Further, scholars argue that both pre-service and in-service teachers must develop a critical consciousness, a state of awareness, activated through dialogue, where one engages in an analysis of context and power (Freire, 1973). Teachers of all disciplines should develop an "engaged pedagogy" that requires a reflective stance that involves interrogating one's location and the identifications and allegiances that inform one's life (hooks, 1994).

There has been a growing body of work that highlights how STEM pre-service and in-service professional development can be designed to achieve the goal of creating reflective, critically conscious teachers who are empowered to address the needs of Black STEM learners and enact pedagogical practices that liberate Black students (Joseph *et al.*, 2016). For example, Johnson (2016), a White mathematics pre-service teacher educator, in a study of mathematics pre-service teachers at a predominately White institution (PWI) engaged her White pre-service teachers in thinking about their "whiteness" and their role in perpetuating stereotypes, and how they could incorporate culturally relevant pedagogy and mathematics for social justice in their classrooms. She cites how many of her White preservice teachers' responses reflected color-blind perspectives, an aversion to "activism" or being viewed as "agents of change" and resistance to seeing themselves as agents of change. As one of her preservice mathematics teachers asserted, "If I had wanted to teach about social issues, I would have become a social studies teacher." Understanding the importance of having her students challenge these notions, she offered a reconceptualization of the definition of activism that she felt could change the way White mathematics teachers see the work of social justice and can, therefore, more easily imagine themselves engaging in interrogating their racial consciousness and seeing themselves as social justice educators (p. 176). To this end, the author offered a reframing of activism to what she called "moment to moment activism," a kind of activism "in which individual people engage in a collection of actions that are intended to promote or direct change within and/or to systems of privilege and oppression" (p. 178). This reframing resulted in her preservice mathematics teachers' willingness to become more reflective and develop the kind of critical consciousness they needed to forward a social justice agenda in their future mathematics teaching.

Other scholars reported ways in which they provided professional development experiences that helped STEM teachers reflect on issues of power, position, and privilege that helped teachers develop the kind of critical consciousness needed to be successful with Black students. These practices included using one's narrative to illuminate how one's racial consciousness has evolved (Harper, 2016), providing experiences that challenge the notion of color blindness and the race-neutral narratives that dominate STEM education (Charles, 2016), forwarding conversations surrounding issues of race and racisms despite the STEM teachers "discomfort" with these topics (Ellington, 2016), and relinquishing notions of "helping the other" that can be viewed as patronizing, self-serving, and "charitable" and adopting ideologies that promote "working with" that can achieve collaboration and partnership and ensure collective agency (Hooks, 1994; Rodriguez, 2016; Spencer, 2016).

Through creating professional development that integrates effective practices in STEM education with practices that promote reflection and critical consciousness, teachers will be more empowered to address the myriad of challenges they face in teaching Black students. Also, they will be better equipped to build the kinds of relationships with these students that reflect a culture of care that will nurture their students' STEM learning, empower them to be resilient learners, and create the classroom environments where Black students feel a sense of belonging. Further, STEM teacher professional development will not only address more than the "struggles" teachers face with teaching STEM content but also address other issues that they face like classroom management, student disengagement, and academic issues from a new perspective that will yield new more positive solutions.

The professional development of STEM teachers must address the need to become competent STEM teachers as well as agents of change for Black students. No matter what form of professional development is used, teachers must heed the message that

> We are more than content experts, we are game changers, agents of transformation, and critical partners in developing our children and cultivating the future STEM doers and scholars. We promote positive STEM identities in our students, ground our work in the lived realities of our students and their communities and are willing to interrogate our own beliefs, perspectives and practices as well as our social contexts to dismantle inequity, social injustice, and racism.

Some considerations that should inform teacher professional development include:

- The ongoing willingness to reflect on one's practice in the context of privilege, power, and positionality
- Engage in activities that connect course content to relationship building and addressing issues of privilege, power, and positionality (historically, socially, politically, and other)
- Being willing to discuss the systemic issues of power, position, and privilege and how these concepts relate to STEM education

- Creating professional communities of practice and affinity groups as "safe space" to discuss diversity-related challenges and opportunities
- Developing practices that help theachers understand the difference between intent and impact, particularly regarding microaggressions and accountability to marginalized communities
- Reframing teachers' role and responsibilities from that of content experts and to agents of change
- Reframing achievement gap discussion to one that focuses on education debt (Ladson-Billings, 2006)

Conclusions

In this chapter, I propose a transformative framework of STEM education that could promote success, persistence, and positive outcomes for Black students and help us achieve equity and diversity outcomes in STEM at all levels. This framework draws on several bodies of work grounded in an assets-view of Black children and reflects research on programs, practices, and perspectives that have been shown to be essential for Black students' success. What makes this framework transformative is that it advocates the integration of various components of Black students' STEM education. The framework moves beyond the curriculum-based reforms that have dominated the STEM landscape to one that is holistic, comprehensive, and integrated. Any program or practice that is adapted to improve the educational outcomes for Black children must first consider who students are, their agency and resiliency, and help them apply these to their STEM learning and help them cultivate the kinds of identities and personal characteristics that foster STEM learning, persistence, and achievement. Also, STEM education that will encourage equity must draw on and contribute to the social and cultural capital of Black students' communities and be done in ways that create and sustain mutually beneficial partnerships between STEM educators and the community. Further, the school-based practices should reflect the transformative STEM-focused leadership that integrates principles of CRT, innovative STEM curricula, instruction, and assessment practices (interdisciplinary, problem-based learning, STEM academies) and social justice and antiracist STEM experiences. Lastly, STEM teachers of Black children require professional development that not only supports them in being competent in STEM teaching and learning, but in being reflective and critically conscious educators who are willing to interrogate their practices in ways that forward a socially just society.

Transformation requires that we not only change what we do but change who we are and the messages we send Black children about the importance of STEM learning. We must become STEM educators who are willing to do more than tweak our practices or incorporate new standards; we must be willing to reform who we are, how we see Black children and their communities, and dismantle our ingrained deficit-based views of Black children. As we reframe who we are and the work we do, we will have the best chance of realizing the diversity and inclusive outcomes we desire.

References

Abrams, D., & Hogg, M. A. (1988). Comments on the motivational status of self-esteem in social identity and intergroup discrimination. *European Journal of Social Psychology, 18*(4), 317–334.

Alvarado, S. E., & Muniz, P. (2015). Do STEM enrichment programs enhance college readiness for racial and ethnic minorities? Conference Papers, American Sociological Association, 1–46.

Andrews, D. J. C. (2012). Black achievers' experiences with racial spotlighting and ignoring in a predominantly white high school. *Teachers College Record, 114*(10), 1–46.

Apple, M. W., Au, W., & Gandin, L. A. (2009). Mapping critical education. In *The Routledge international handbook of critical education* (pp. 3–19) New York: Routledge.

Asante, M. K. (1991). The Afrocentric idea in education. *The Journal of Negro Education, 60*(2), 170–180.

Asghar, A., Ellington, R., Rice, E., Johnson, F., & Prime, G. M. (2012). Supporting STEM education in secondary science contexts. *Interdisciplinary Journal of Problem-based Learning, 6*(2), 4.

Bass, L. (2012). When care trumps justice: The operationalization of Black feminist caring in educational leadership. *International Journal of Qualitative Studies in Education, 25*, 73–87.

Baumeister, R. F., & Leary, M. R. (1995). The need to belong: Desire for interpersonal attachments as a fundamental human motivation. *Psychological Bulletin, 117*(3), 497.

Bergin, D. A., & Cooks, H. C. (2000). Academic competition among students of color: An interview study. *Urban Education, 35*(4), 442–472.

Bergin, D. A., & Cooks, H. C. (2002). High school students of color talk about accusations of "acting White." *The Urban Review, 34*(2), 113–134.

Berry, R. Q., III (2003). Mathematics standards, cultural styles, and learning preferences: The plight and the promise of African American students. *The Clearing House, 76*(5), 244–249.

Brown, B. A., Mangram, C., Sun, K., Cross, K., & Raab, E. (2017). Representing racial identity: Identity, race, and the construction of the African American STEM students. *Urban Education, 52*(2), 170–206.

Bryk, A. S., Sebring, P. B., Allensworth, E., Luppescu, S., & Easton, J. Q. (2010). *Organizing schools for improvement: Lessons from Chicago.* Chicago, IL: The University of Chicago Press.

Charles, R. (2016). Response to teacher educators and pre-service teachers working through complexities of whiteness and race in mathematics education. In N. M. Joseph, C. Haynes, & F. Cobb (Eds.), *Interrogating whiteness and relinquishing power: White faculty's commitment to racial consciousness in STEM classrooms*, (pp. 191–210). New York, NY: Peter Lang.

Cheryan, S., Plaut, V. C., Davies, P. G., & Steele, C. M. (2009). Ambient belonging: How stereotypical cues impact gender participation in computer science. *Journal of Personality and Social Psychology, 97*(6), 1045.

Coleman, J. S. (1990). *Foundations of social theory.* Cambridge: Harvard University Press.

Coleman, J. S. (2000). Social capital in the creation of human capital. In *Knowledge and social capital* Butterworth Heinemann: Boston (pp 17–41).

Cook, L. (2014). *Mentor/mentee relationships: The experience of African American STEM majors* (Doctoral dissertation). Morgan State University.

Desimone, L. M. (2009). Improving impact studies of teachers' professional development: Toward better conceptualizations and measures. *Educational Researcher, 38*(3), 181–199.

DiMaggio, P. (1982). Cultural capital and school success: The impact of status culture participation on the grades of U.S. high school students. *American Sociological Review, 47*(2), 189–201. Retrieved from http://www.jstor.org/stable/2094962

Dimick, A. S. (2012). Student empowerment in an environmental science classroom: Toward a framework for social justice science education. *Science Education, 96*(6), 990–1012.

Duckworth, A. L., Peterson, C., Matthews, M. D., & Kelly, D. R. (2007). Grit: Perseverance and passion for long-term goals. *Journal of Personality and Social Psychology, 92*(6), 1087.

Ellington, R. (2006). *Having their say: Eight high-achieving African-American undergraduate mathematics majors discuss their success and persistence in mathematics* University of Maryland College Park, College Park, MD (Doctoral dissertation).

Ellington, R. M. (2016). Mathematics teacher education as racialized experience: One black scholar's response to a white teacher educator's critical consciousness evolution and social justice practice. In N. M. Joseph, C. Haynes, & F. Cobb (Eds.), *Interrogating whiteness and relinquishing power: White faculty's commitment to racial consciousness in STEM classrooms* (pp. 211–222). New York, NY: Peter Lang.

Ellington, R. M., & Frederick, R. (2010). Black high achieving undergraduate mathematics majors discuss success and persistence in mathematics. *Negro Educational Review, 61*(1–4), 61.

Ford, M. R. (2018). Approaches to school leadership in inclusive STEM high schools: A cross-case analysis. *Dissertation Abstracts International Section A, 78*.

Franklin, V. P. (Ed.). (2004). *Cultural capital and black education: African American communities and the funding of black.* Charlotte, NC: IAP.

Freire, P. (1973). *Education for critical consciousness.* New York, NY: Seabury.

Fries-Britt, S. L., Younger, T. K., & Hall, W. D. (2010). Lessons from high-achieving students of color in physics. *New Directions for Institutional Research, 2010*(148), 75–83.

Gay, G. (2000). *Culturally responsive teaching: Theory, research, and practice* (2nd ed.). New York, NY: Teachers College Press.

Gay, G. (2002). Preparing for culturally responsive teaching. *Journal of Teacher Education, 53*(2), 106–116.

Gay, G., & Kirkland, K. (2003). Developing cultural critical consciousness and self-reflection in preservice teacher education. *Theory into Practice, 42*(3), 181–187.

Geertz, C. (2003). Racial gaps found to persist in public's opinion of schools. *Education Week on the Web.* Retrieved from http://www.edweek.org/ ew/newstory.cfm?slug=37race.h22

Gergen, K. J. (1991). *The saturated self: Dilemmas of identity in contemporary life.* New York, NY: Basic.

Getz, C. (2000). Observing the spirit of resilience: The relationship between life experiences and success in higher education for African-American students. In S. Gregory (Ed.), *The academic achievement of minority students: Perspectives, practices, and prescriptions* (pp. 457–490). New York, NY: University Press of America.

Goodwin, L. (2002). *Resilient spirits: Disadvantaged students making it at an elite university.* New York, NY: Routledge Falmer.

Green, M., Emery, A., Sanders, M., & Anderman, L. H. (2016). Another path to belonging: A case study of middle school students' perspectives. *Educational & Developmental Psychologist, 33*(1), 85. doi:10.1017/edp.2016.4.

Grigg, J., Kelly, K. A., Gamoran, A., & Borman, G. D. (2013). Effects of two scientific inquiry professional development interventions on teaching practice. *Educational Evaluation and Policy Analysis, 35*(1), 38–56.

Gutiérrez, R. (2008). A "gap-gazing" fetish in mathematics education? Problematizing research on the achievement gap. *Journal for Research in Mathematics Education, 39*(4), 357–364.

Gutstein, E. (2006). *Reading and writing the world with mathematics: Toward a pedagogy for social justice*. London: Taylor & Francis.

Gutstein, E. (2007). Connecting community, critical, and classical knowledge in teaching mathematics for social justice (pp. 109–118). The Montana Mathematics Enthusiast, Monograph, 1.

Harper, F. K. (2016). Challenging patterns to change my world: Using my personal evolution of critical race consciousness in mathematics teacher education. In N. M. Joseph C. Haynes, & F. Cobb (Eds.), *Interrogating whiteness and relinquishing power: White faculty's commitment to racial consciousness in STEM classrooms* (pp. 152–169). New York, NY: Peter Lang.

Harper, S. R. (2010). An anti-deficit achievement framework for research on students of color in STEM. *New Directions for Institutional Research, 2010*(148), 63–74. Professional Development Collection, EBSCOhost (accessed February 17, 2018).

hooks, b. (1994). *Teaching to transgress: Education as the practice of freedom*. New York, NY: Routledge.

Howard, T. C. (2003). Culturally relevant pedagogy: Ingredients for critical teacher reflection. *Theory into Practice, 42*(3), 195–202.

Huberman, M. (1991). The pedagogy of poverty versus good teaching. *Phi Delta Kappan, 73*(4), 290–294.

Hubert, T. L. (2013). Learners of mathematics: High school students' perspectives of culturally relevant mathematics pedagogy. *Journal of African American Studies, 18*, 324–336. doi:10.1007/s12111-013-9273-2.

Jencks, C., & Phillips, M. (Eds.). (2011). *The black-white test score gap*. Washington, DC: Brookings Institution Press.

Jett, C. C. (2009). Mathematics, an empowering tool of liberation? A review of mathematics teaching, learning, and liberation in the lives of black children. *Journal of Urban Mathematics Education, 2*(2), 66–71.

Johnson, D. R. (2012). Campus racial climate perceptions and overall sense of belonging among racially diverse women in STEM majors. *Journal of College Student Development, 53*(2), 336–346.

Johnson, K. (2016). Reconceptualizing "activism": Developing a socially conscious practice with prospective white mathematics teachers. In N. M. Joseph, C. Haynes, & F. Cobb (Eds.), *Interrogating whiteness and relinquishing power: White faculty's commitment to racial consciousness in STEM classrooms* (pp. 171–187). New York, NY: Peter Lang.

Johnson, M. L. (1984). Blacks in mathematics: A status report. *Journal for Research in Mathematics Education, 15*(2), 145–153.

Joseph, N. M., Haynes, C., & Cobb, F. (Eds.). (2016). *Interrogating whiteness and relinquishing power: White faculty's commitment to racial consciousness in STEM classrooms*. New York, NY: Peter Lang.

Kane, J. M. (2012). Young African American children constructing academic and disciplinary identities in an urban science classroom. *Science Education, 96*(3), 457–487.

Ladson-Billings, G. (1995a). But that's just good teaching! The case for culturally relevant pedagogy. *Theory into Practice, 34*, 159–165.

Ladson-Billings, G. (1995b). Toward a theory of culturally relevant pedagogy. American *Educational Research Journal, 32*(3), 465–491.

Ladson-Billings, G. (2006). From the achievement gap to the education debt: Understanding achievement in US schools. *Educational Researcher, 35*(7), 3–12.

Laughter, J. C., & Adams, A. (2012). Culturally relevant science teaching in middle school. *Urban Education, 47*, 1104–1132.

Laura, B., McMeeking, S., Orsi, R., & Cobb, R. B. (2012). Effects of a teacher professional development program on the mathematics achievement of middle school students. *Journal for Research in Mathematics Education, 43*(2), 159–181.

Lee, A. (2015). Determining the effects of computer science education at the secondary level on STEM major choices in postsecondary institutions in the United States. *Computers & Education, 88*, 241–255. doi:10.1016/j.compedu.2015.04.019

Leonard, J. (2017). *Culturally specific pedagogy in the mathematics classroom: Strategies for teachers and students.* London: Routledge.

Leonard, J., Brooks, W., Barnes-Johnson, J., & Berry III, R. Q. (2010). The nuances and complexities of teaching mathematics for cultural relevance and social justice. *Journal of Teacher Education, 61*(3), 261–270.

Leonard, J., & Martin, D. B. (Eds.). (2013). *The brilliance of Black children in mathematics.* Charlotte, NC: IAP.

Lin, N. (2001). *Social capital: A theory of social structure and action.* Cambridge: Cambridge University Press.

Long III, L. L., & Henderson, T. S. (2017). Lessons learned from successful black male "Buoyant Believers" in engineering and engineering-related fields. *Proceedings of the ASEE Annual Conference & Exposition*, Embry-Riddle Aeronautical University, Columbus, OH1.

Lord-Walker, J. (2015). *Highly successful outcomes: How teachers at an African-centered independent school structure mathematics curriculum and instruction* (Doctoral dissertation), Mills College.

Loucks-Horsley, S., Hewson, P., Love, N., & Stiles, K. (1998). Ideas that work: Mathematics professional development. The Eisenhower National Clearinghouse for Mathematics and Science Education, Washington, DC.

Lubienski, S. T. (2002). A closer look at black-white mathematics gaps: Intersections of race and SES in NAEP achievement and instructional practices data. *Journal of Negro Education, 71*(4)269–287.

Martin, D. B. (2010, March). Not-so-strange bedfellows: Racial projects and the mathematics education enterprise. In *Proceedings of the sixth international mathematics education and society conference* (Vol. 1, pp. 57–79). Berlin: Freire University Berlin.

Martin, D. B. (2012). Learning mathematics while black. *The Journal of Educational Foundations, 26*(1/2), 47.

Martin, D. B., Gholson, M. L., & Leonard, J. (2010). Mathematics as gatekeeper: Power and privilege in the production of knowledge. *Journal of Urban Mathematics Education, 3*(2), 12–24.

McGee, E. O. (2015). Robust and fragile mathematical identities: A framework for exploring racialized experiences and high achievement among black college students. *Journal for Research in Mathematics Education, 46*(5), 599–625.

McGee, E. O., & Bentley, L. (2017). The troubled success of black women in STEM. *Cognition and Instruction, 35*(4), 265–289.

McGee, E. O., & Martin, D. B. (2011). "You would not believe what I have to go through to prove my intellectual value!" Stereotype management among academically successful Black mathematics and engineering students. *American Educational Research Journal, 48*(6), 1347–1389.

Mihyeon, K., Cross, J., & Cross, T. (2017). Program development for disadvantaged high-ability students. *Gifted Child Today, 40*(2), 87.

Milner IV, H. R. (2013). Analyzing poverty, learning, and teaching through a critical race theory lens. *Review of Research in Education, 37*(1), 1–53.

Montgomery, W. (2001). Creating culturally responsive, inclusive classrooms. *Teaching Exceptional Children, 33*(4), 4–9.

Moody, V. (2001). The social constructs of the mathematical experiences of African-American students. Sociocultural research on mathematics education, 255–278.

Moody, V. R. (1997). *Giving voice to African Americans who have been successful with school mathematics* (Unpublished doctoral dissertation). The University of Georgia, Athens.

Moody, V. R. (2000). African American students' success with school mathematics. In M. E. Strutchens, M. L. Johnson, & W. F. Tate (Eds.), *Changing the faces of mathematics: Perspectives on African Americans* (pp. 51–60). Reston, VA: National Council of Teachers of Mathematics.

Moody, V. R. (2004). Sociocultural orientations and the mathematical success of African American students. *The Journal of Educational Research, 97*(3), 135–146.

Murrell, P. C., Jr. (2012). *African-centered pedagogy: Developing schools of achievement for African American children*. Albany, NY: SUNY Press.

Nasir, N. I. S., & Vakil, S. (2017). STEM-focused academies in urban schools: Tensions and possibilities. *Journal of the Learning Sciences, 26*(3), 376–406.

Nieto, S. (2006). Solidarity, courage, and heart: What teacher educators can learn from a new generation of teachers. *Intercultural Education, 17*(5), 457–473.

Nieto, S. P., Kang, B. E., & Raible, J. (2008). Nieto, Sonia, Patty Bode, Eugenie Kang, and John Raible," Identity, Community, and Diversity: Retheorizing Multicultural Curriculum for the Postmodern Era," pp. 176–197 in F. Michael Connelly, Ming Fang He, and JoAnn Phillion, eds., The SAGE Handbook of Curriculum and Instruction. Los Angeles: Sage Publications, 2008.

Oakes, J. (1990). Opportunities, achievement, and choice: Women and minority students in science and mathematics. In C. B. Cazden (Ed.), *Review and research in education* (Vol. 16, pp. 153–222). Washington, DC. American Educational Research Association.

Ogbu, J. U. (1978). Black-white differences Ogbu, J. U. (1978). Black-White differences in school performance: A critique of current explanations. Minority education and caste: The American system in cross-cultural perspective, 43–65.

Ogbu, J. U. (2003). *Black American students in an affluent suburb: A study of academic disengagement*. Mahwah, NJ: Erlbaum.

Oyserman, D. (2007). Social identity and self-regulation. In A. W. Kruglanski & E. T. Higgins (Eds.), *Social psychology: Handbook of basic principles* (2nd ed., pp. 432–453). New York, NY: Guilford Press.

Oyserman, D. (2008). Racial-ethnic self-schemas: Multi-dimensional identity-based motivation. *Journal of Research on Personality, 42*, 1186–1198.

Oyserman, D. (2009). Identity-based motivation: Implications for action-readiness, procedural readiness, and consumer behavior. *Journal of Consumer Psychology, 19*, 250–260.

Oyserman, D., Brickman, D., & Rhodes, M. (2007). Racial-ethnic identity in adolescence: Content and consequences for African American and Latino and Latina youth. In A. Fuligni (Ed.), *Contesting stereotypes and creating identities: Social categories, identities and educational participation* (pp. 91–114). New York, NY: Russell Sage Foundation.

Robinson, V. M. J., Lloyd, C. A., & Rowe, K. J. (2008). The impact of leadership on student outcomes: An analysis of the differential effects of leadership types. *Educational Administration Quarterly, 44*, 635–674.

Rodriguez, A. (2016). For whom do we do equity and social justice work? Recasting the discourse about the other to effect transformative change. In N. M. Joseph, C. Haynes, & F. Cobb (Eds.), *Interrogating whiteness and relinquishing power: White faculty's commitment to racial consciousness in STEM classrooms*. New York, NY: Peter Lang.

Rodriguez, J. L., Bustamante Jones, E., Peng, V. O., & Park, C. D. (2004). Promoting academic achievement and identity development among diverse high school students. *High School Journal, 87*, 44–53. doi:10.1353/hsj.2004.0002.

Sandoval-Lucero, E., Maes, J., & Klingsmith, L. (2014). African American and Latina (o) community college students' social capital and student success. *College Student Journal, 48*(3), 522–533.

Shields, C. (2012). *Transformative leadership in education: Equitable change in an uncertain and complex world*. New York, NY: Routledge.

Shockley, K. G. (2007). Literatures and definitions: Toward understanding Afrocentric education. *The Journal of Negro Education, 76*(2) 103–117.

Shockley, K. G., & Cleveland, D. (2011). Culture, power, and education: The philosophies and pedagogy of African centered educators. *The International Journal of Critical Pedagogy, 3*(3) pp 54–75.

Smith, W. A., Allen, W. R., & Danley, L. L. (2007). "Assume the position... you fit the description" psychosocial experiences and racial battle fatigue among African American male college students. *American Behavioral Scientist, 51*(4), 551–578.

Spencer, J. (2016). Moving from the outside in or what white colleagues need to do to get it right with their white students In N. M. Joseph, C. Haynes, & F. Cobb (Eds.), *Interrogating whiteness and relinquishing power: White faculty's commitment to racial consciousness in STEM classrooms*. New York, NY: Peter Lang.

Stanton-Salazar, R. D., & Spina, S. U. (2000). The network orientations of highly resilient urban minority youth: A network-analytic account of minority socialization and its educational implications. *Urban Review, 32*(3), 227–261.

Stinson, D. W. (2006). African American male adolescents, schooling (and mathematics): Deficiency, rejection, and achievement. *Review of Educational Research, 76*(4), 477–506.

Strayhorn, T. L. (2015). Factors influencing black males' preparation for college and success in STEM majors: A mixed methods study. *Western Journal of Black Studies, 39*(1), 45.

Thirutnurthy, V., Kirylo, J. D., & Ciabattari, T. (2010). Issue in education: Cultural capital, social capital, and educational inequality. *Childhood Education, 87*(2), 119–121.

Thompson, L., & Davis, J. (2013). The meaning high-achieving African-American males in an urban high school ascribe to mathematics. *The Urban Review, 45*(4), 490–517.

Vakil, S. (2014). A critical pedagogy approach for engaging urban youth in mobile app development in an after-school program. *Equity and Excellence in Education, 47*(1), 31–45.

Vanneman, A., Hamilton, L., Anderson, J. B., & Rahman, T. (2009). Achievement gaps: How black and white students in public schools perform in mathematics and reading on the National Assessment of Educational Progress. Statistical Analysis Report. NCES 2009–455. National Center for Education Statistics.

Varelas, M., Martin, D. B., & Kane, J. M. (2012). Content learning and identity construction: A framework to strengthen African American students' mathematics and science learning in urban elementary schools. *Human Development, 55*(5–6), 319–339.

Ward, S. C., Bagley, C., Lumby, J., Woods, P., Hamilton, T., & Roberts, A. (2015). School leadership for equity: Lessons from the literature. *International Journal of Inclusive Education, 19*(4), 333–346.

Wilson, C. M. (2016a). Enacting critical care and transformative leadership in schools highly impacted by poverty: An African-American principal's counter-narrative. *International Journal of Leadership in Education, 19*(5), 557–577.

Wilson, C. M., Douglas, T. M. O., & Nganga, C. (2013). Starting with African American success: A strengths-based approach to transformative educational leadership. In *Handbook of research on educational leadership for equity and diversity*. New York: Routledge (pp. 111–133).

Wilson, R. S. (2016b). *Understanding the experiences and perceptions of African American female STEM majors at a single-sex HBCU* (Doctoral dissertation). Morgan State University.

Wojnowski, B. S., & Pea, C. H. (Eds.). (2014). *Models and approaches to STEM professional development*. Arlington, VA: NSTA Press.

Yosso, T. J. (2005). Whose culture has capital? A critical race theory discussion of community cultural wealth. *Race, Ethnicity and Education, 8*(1), 69–91.

Young, J. R., Young, J. V., & Hamilton, C. (2013). Culturally relevant project-based learning for STEM education: Implications and examples for urban schools. In *Improving urban schools: Equity and access in K-16 STEM education* (pp. 39–65) Charlotte NC: Information Age.

Zimmerman, B. J., & Cleary, T. J. (2006). Adolescents' development of personal agency: The role of self-efficacy beliefs and self-regulatory skill. *Self-efficacy Beliefs of Adolescents, 5,* 45–69.

Reconceptualizing Science Education for Learners of African Descent

JOMO W. MUTEGI, CRYSTAL H. MORTON, AND LESLIE K. ETIENNE

Abstract

This chapter is a description of practice grounded in the idea that the primary problems Black children face in schools are political problems. The chapter articulates three aspects of science education that should be reconceptualized if we are to adequately address these problems. These three aspects are: the purpose of science education; science content; and the role of the instructor. The theoretical foundation for reconceptualizing these three aspects of science education comes from Goduka's (2005) articulation of e*Ziko*, Mutegi's (2011) articulation of socially transformative STEM curriculum, and Codrington's (2014) work on liberatory education. Drawing from this theoretical foundation, the chapter illustrates the how science educators could reconceptualize the purpose of science education, science content, and the role of the instructor by describing a year-long project in which three, high school-aged, young ladies and one university professor worked collaboratively as science writers. Through the Black Kids Read - Science Writers project, these young ladies took on the task of authoring science-oriented literature for elementary-aged children.

Introduction

Amos Wilson once said, "The problems that Black children face in schools are not cognitive problems, neither are they pedagogical problems. The problems that Black children face in schools are political problems."

Given the sociopolitical history of African people, Wilson's words might seem self-evident. However, the political nature of the problems Black children face in schools is barely discernable in science education research literature. Studies of the science education of African American learners focus inordinately on disparities between African American learners and their peers (e.g., Cohen, Garcia, Purdie-Vaughns, Apfel, & Brzustoski, 2009; Muller, Stage, & Kinzie, 2001; Norman, Ault, Bentz, & Meskimen, 2001). What is more, these studies rarely explain the schooling of African American students in terms of colonialism, enslavement, Jim Crow segregation, state sanctioned killings, mass incarceration, or other sociopolitical structures aimed at suppressing people of African descent. Instead, these studies explain disparity by claiming deficiencies in African American students, their culture and communities (Lewis, Pitts, & Collins, 2002).

In terms of practice, the political nature of the problems Black children face in schools is completely absent in science education literature written for science teachers. In fact, the Blackness of children themselves is completely absent. Mutegi and Morton (2012) point out that for the 25 years between 1984 and 2011 only 8 of over 7,000 articles (0.11%) in NSTA practitioner journals[1] address African Americans in any way. Only 2 of these 8 articles describe pedagogical approaches that are best suited for African American learners.

This chapter aims to fill the void in existing literature and respond to the issue Wilson raised by articulating three aspects of science education that should be reconceptualized in order to address the political nature of the problems faced by science learners of African descent. These are: *reconceptualizing the purpose of science education, reconceptualizing science content,* and *reconceptualizing the role of the instructor.*

The theoretical foundation for the reconceptualization described in the chapter comes from Goduka's theoretical framework for indigenous knowledge systems (Goduka, Madolo, Rozani, Notsi, & Talen, 2013), Mutegi's (2011) articulation of socially transformative STEM curriculum, and Codrington's (2014) work on liberatory education. Drawing from these theoretical foundations, the chapter describes a year-long project in which three, high school-aged, young ladies and one university professor worked collaboratively as science writers.

Theoretical Framework

eZiko siPheka siSophula (*eZiko* for short) is a theoretical framework presented by Goduka (2005) and has been applied to both teaching (Goduka *et al.*, 2013; Mashoko, 2014) and sustainable development (Taleni & Goduka, 2013) in African contexts. Rooted in the Nguni language, the phrase is translated as "gathering around the hearth (iziko) to cook (sipheka) and dish out (sisophula)." In Nguni life, the eZiko (or the hearth in this instance) is physically located in the center of the mud hut homes in which people live. eZiko is also a metonym for the village as a central location in a person's life cycle. As an example, it is common practice among the Xhosa for one's village to be the place where one's umbilical cord is buried; and upon death one's body must be returned to that same village.

In Goduka's framework, *eZiko siPheka siSophula* is a metaphor for praxis that is feminine; experiential and participatory; and collective and communal. The characteristics of the praxis that are feminine are those that emphasize qualities of caring and nurturing. Considering the translation "gathering around the hearth to cook and dish out" the framework reminds us of the caring and nurturing aspects of being around a fire and providing sustenance (cooking and dishing out). The characteristics of the praxis that are experiential and participatory are those which emphasize the degree to which all participants are actively engaged in the work that is under study. The characteristics of the praxis that are collective and communal are those that emphasize the positionality of the participants. This is indicated by the prefix "si" which means "we cook" as opposed to "ndi" which might mean "I cook." So, siPheka (we cook) and siSophula (we dish out) emphasize that all participants have a vested interest in the praxis. Taken together, these three characteristics of *eZiko siPheka siSophula* represent an approach to praxis wherein an entire community actively engages in work aimed at enhancing (or nurturing) that community.

Socially Transformative STEM Curriculum is a curricular approach developed in response to the sociocultural positioning of people of African descent worldwide. In presenting this approach, Mutegi (2011) argues that African people continue to live a colonized, slave-like existence. In response to this, their education should be one that positions them to dismantle systemic racism. Building on the work of critical theorists (e.g., Allen, 2004; Freire, 1970; Macedo, 1993), Mutegi and colleagues have applied the notion of socially transformative curriculum to both science (Mutegi, Lewis, & Smith-Mutegi, 2017; Mutegi & Morton, 2012) and mathematics (Pitts Bannister, Davis, Mutegi, Thompson, & Lewis, 2017).

Their approach argues that STEM curricula should position learners to attain five types of mastery. These are: *content*, which is mastery of the traditional STEM content; *currency*, which is mastery of how the content is relevant to human beings broadly; *context*, which is mastery of how the content is specifically relevant to people of African descent; *critique*, which is mastery of how knowledge of the content helps us to better understand how systemic racism is established and maintained; and *conduct*, which is mastery of some readily implementable skill or set of skills that position students to apply their understanding in ways that disrupts and ultimately dismantles systemic racism.

Codrington (2014) pens an essay in which she calls for a more focused treatment of culturally responsive science teaching, one that moves toward liberatory education. At the heart of Codrington's argument is the idea that "...education must tend to one's paramount needs, and that the paramount need for African Americans—whose history is marked by oppression—is liberation" (p. 1018). In an effort to empower science educators, Codrington offers questions on which science educators can reflect as they work to move toward more liberatory practices. Among the questions she offers are:

- How are most systems of science epistemologically and ontologically hegemonic in nature, continuing to perpetuate cycles of racism, colonialism, and Western values steeped in dominance over people and consumption of natural resources?
- How can science be placed in a historical and cultural context so that science educators do not perpetuate the notion that scientists come from predominately Western, middle-class White cultures?
- How do the commonly held views of science embody racial stereotypes?
- What impact do racial stereotypes have on the learning of science and motivation to practice science as a profession?
- How can the practice of science be deconstructed to demonstrate its political nature both historically and in contemporary times, and debunk the myth of natural science as universal, objective, and value-free?

The work of Goduka, Mutegi, and Codrington forms the theoretical foundation on which the following description of practice is based. Throughout the description, we will take time to illustrate for the reader instances where the implemented practice and the theoretical foundation complement one another.

Reconceptualizing Science Education

As previously stated, the aim of this chapter is to articulate three aspects of science education that should be reconceptualized in order to address the political nature of the problems faced by science learners of African descent. To illustrate how these aspects might look in practice, we provide a detailed description of an informal science education effort called Black Kids Read—Science Writers Project. Throughout this description, we will also revisit elements of the previously articulated theoretical foundation in order to demonstrate the complementarity between the theory and practice described.

Context

Black Kids Read is an effort started by the first author (Mutegi) to produce science-related books for Black children. The effort began in part to (a) address the dearth of Black children's literature (Myers, 2014; Myers & Myers, 2014) and (b) provide materials that parents and teachers could use to support the science learning of Black children. Black Kids Read books profile noted Black scientists (e.g., Mutegi, 2016); retell science-related historical events from Black perspectives (e.g., Mutegi, 2015b); and feature Black protagonists, living in Black communities, using science to solve problems facing those communities (e.g., Mutegi, 2015a).

The Black Kids Read—Science Writers Project is an extension of that effort. In the first year of the Science Writers Project, three high school-aged young ladies were hired to work with Mutegi as science writers. Through their work as science writers they learned to author science-related texts (i.e., news articles and short stories) intended for use by elementary-aged African American children.

Although the Black Kids Read—Science Writers Project was not designed to service adolescent females exclusively, the young ladies who participated were recruited from a STEM summer camp called Girls STEM Institute (Morton, 2017; Morton & Smith-Mutegi, 2017) run by the second author (Morton). All high school-aged young ladies in the camp were invited to apply to become science writers by submitting a letter of interest, a resume, and one letter of reference. Between the months of September and June, students hired to work as science writers worked on campus one day a week for about 2.5 hours each day. Although science writers often had additional work that they could complete at home throughout the week, they were paid hourly only for the time they were on campus.

Reconceptualizing the Purpose of Science Education

Reconceptualizing the purpose of science education requires a shift from a state-sponsored effort to bolster the workforce and consumer base toward an effort to dismantle systemic racism.

Description of Practice

The Science Writers (Daisha, Antoinette, Kiona, and Baba Mutegi) sat in rapt attention as the narrator continued,

> Against the broad spectrum of time-worn caricatures, the reality of Black life in the early 1900s was undergoing dramatic change. In growing numbers, Blacks were moving from the country to the city, from the South to the North. Emancipation has disrupted the social order of the South; now Black migration and competition for jobs threatened the status quo of the North. Racial hostilities began to brew. New caricatures of the urban coon emerged, reflecting the perceived threat of an expanding Black labor force.

Daisha motioned for the documentary to be paused. She then shared a thought with the group, "So, it looks like all of the stereotypes are made to put emphasis on the negative aspects of Black people."

Antoinette responded, "Well that is only partly true. The stereotypes are creating negative aspects of Black people. Our lips don't look anything like that, and the behavior they are describing is not our behavior." The science writers spent five minutes or so unpacking the purpose and the impact of the Zip Coon archetype before returning to the documentary.

The documentary *Ethnic Notions* (Riggs, 1987) profiles six archetypes that whites created to misrepresent Black people in media. These archetypes are not presented without context. Instead, the documentary shows what historical events precipitated white's creation of each archetype; what the archetypes are intended to communicate to media consumers; and to a lesser degree, the impact of these archetypes on the public perception of African people and their Africanness. Following the viewing and discussion of the documentary, the science writers explored whether and to what degree these archetypes were still present in modern media. By the end of this brainstorming discussion the science writers generated a list of fairly contemporary media where one or more of these archetypes could be found.

"Archetype is only one element of media." Baba Mutegi continued, "The other three are: plot, narrative, and…".

Antoinette interrupted excitedly, "Oh! We learned this in English class. I can't remember the fourth one."

Kiona coolly chimed in, "Theme."

"Yeah! Theme."

Together, Baba Mutegi and the science writers explored the differences between these four elements of media. The science writers drew heavily on what they had learned from their English class. Baba Mutegi pushed them to apply their understanding to a wide variety of media portrayals of people of African descent. Throughout the discussion they drew examples from a broad range of media sources, including: books (both fiction and non-fiction), newspapers and magazines (stories, advertisements, and pictorial representations), and video media (such as movies, television programs, and video games).

The science writers also discussed the degree to which popular media portrayals of Black people (as evidenced in plots, archetypes, narratives, and themes) are accurate reflections of Black people that "we know in our daily lives." The viewing of *Ethnic Notions* served as the foundation of a three-week exploration into the structure and function of media as it pertains to people of African descent. Throughout this three-week exploration, the science writers examined a wide range of media, critiqued and discussed the documentary, rewatched and discussed portions of the documentary with their families, and worked to use the elements of media to critique and better understand contemporary media portrayals. Such an intense focus on media and media representation rests outside of the stated purpose of science education.

This three-week exploration culminated in an exercise where science writers studied the work of an African American physicist, Dr. Hadiya Nicole Green, and began the process of planning a text that would retell the story of her life and work. As of this writing, Dr. Green is a professor in the Morehouse School of Medicine who is pioneering research in the use of nanotechnology to provide targeted treatment of cancer cells. According to media accounts, Dr. Green's interest in cancer research was fostered in part by her experience of caring for both an aunt and uncle as they died of cancer. The science writers chose to explore and retell Dr. Green's story over that of other possible scientists. They were moved in part by the challenges that Dr. Green faced early in life and they were also intrigued to find an African American female with a youthful demeanor who was an accomplished scientist.

"Stories do not write themselves...even stories about the lives of scientists. So, what do we want the world to know about Dr. Hadiyah-Nicole Green."

For two weeks the science writers read news stories, watched media interviews, and examined university websites to learn as much as possible about Dr. Green. We took time in our weekly meetings to share what we found and to piece together disparate bits of information into a cohesive story. We also argued back and forth as we worked to fill in the gaps. What conclusions could we draw about her personality without a direct report? What assertions would be reasonable? What type of person might she be? What type of people might her aunt and uncle have been?

As we neared the end of the second week and we began to feel that we had a reasonably good picture, we began outlining a plot for our story. Kiona spoke up eagerly, "I've outline 3 narratives that I think we should work to include." She then shared her narratives.

Antoinette protested, "We can't just make up events that didn't happen."

Smiling, Daisha pushed the issue, "Why can't we? Everything we have looked at so far is people making up things...events, characters, motives. Even the news makes up things. Even when you try to tell a story accurately, you tell it from your perspective. So in a way, you are making up a story."

Baba Mutegi asked, "Antoinette, how would you tell this story?"

The science writers each told a different version of the story they outlined and provided peer critique to one another. As science writers each retold their version of Dr. Green's story, we realized that each retelling focused more on different parts of her life than others. One focused more on her emotional response to life events than others. One focused more on her cognition and commitment to academics than others. One focused more on the social circumstances that provided the backdrop for her life events. They questioned one another to better understand each other's respective versions of the story. Through this process, Antoinette saw that there is some element of fiction in every story. They then agreed to follow Kiona's lead, and they identified a set of archetypes, narratives and themes to drive their plot. They also agreed that these archetypes, narratives and themes should (a) reflect Black people that "we know in our daily lives," and (b) positively shape the image of African people.

The story of Hadiyah-Nicole Green, served as a culminating application of an extensive focus on media. This is one example of how the

Science Writers Project reconceptualizes the purpose of science education. For the science writers, the primary purpose of science education as represented in this episode is not to develop an understanding of science content (or even scientific inquiry) that mirrors the understanding held by scientists. The primary purpose of science education as represented in this episode is to understand how media can be used to either empower (or disempower) African people in general and in science in particular.

So, an intense focus on elements of media and media portrayal is essential if science writers are to be (a) critical consumers of media which often misrepresents the role of science in the history of African people, and (b) change agents as they work to represent science in the lives of African people in empowering ways.

Theoretical Grounding

The practice described here reflects elements of all three theoretical perspectives. It reflects Goduka's (2005) notion of eZiko in that it is caring and nurturing. Rather than doggedly adhering to mandated content, the science writers focus on content that provides sustenance (or that meets the needs of the learners). It also reflects Codrington's (2014) focus on liberatory practice in that it tends to students' "paramount need," which is liberation. Finally, it reflects Mutegi's (2011) articulation of socially transformative curriculum in that the ultimate purpose driving the curriculum is that of positioning science writers to recognize systemic racism and positioning them to dismantle it. This purpose is reflected in each of the five areas of mastery. Science writers explore *content* as they examine the life and work of Hadiyah Nicole Green. They explore *currency* as they examine the mechanisms by which media is produced. They explore *context* as they learn the historical development of archetypes used to represent people of African descent. They explore *critique* as they examine the various ways that media has been used to disadvantage people of African descent; and they explore *conduct* as they work to develop empowering counternarratives that better represent the life and work of a noted African American scientist.

Reconceptualizing Science Content

Reconceptualizing science content requires a shift from an emphasis on standards-mandated content toward an emphasis on content deemed meaningful to students and teachers.

Description of Practice

Baba Mutegi realized that he had made an error in judgment. He had expected that he and the science writers could use a children's picture book as a vehicle for learning about writing. Even though a picture book can be much shorter and simpler than a novel, he found that it was too long and complex for the science writers to use it for immediate feedback on their progress. He also realized that a joint writing project made it more difficult for science writers to foreground individual strengths and challenges. "I would like for us to shift gears a bit. I would like for us to take a few weeks where we can each write a magazine-length article on a science-related topic of our choice. Let's also plan for the article to be read by children in grades 4, 5, and 6."

Daisha wanted to confirm what she had heard, "The article can be on any topic?"

Baba Mutegi replied, "Yes. Any science-related topic."

Daisha continued, "Can I do mine on animal extinction?"

"Yes."

Antoinette also wanted to confirm what she had heard, "Is hair a science-related topic?"

Baba Mutegi replied, "It can be."

Antoinette continued, "I want to write an article on why some people have curly hair and why some people have straight hair."

The project that Mutegi thought would take only a few weeks, actually took 3 months. Each science writer began by reading more about and taking reference notes on her chosen topic. Science writers then began writing drafts of their articles. Once all science writers had a complete draft, each shared her paper with the group and received peer feedback.

After reading Daisha's article on lion extinction, Kiona challenged Daisha's take on the cause of lion extinction. "You write that humans are causing lion extinction through excessive hunting and habitat loss. I think that is misleading."

Daisha defended her position, "How is it misleading? Who or what else would you attribute hunting and habitat loss to if not humans?"

Kiona responded confidently, "If Antoinette bullied you every time you came here, you could say that 'people' were bullying you. You could say that the science writers were bullying you. But, both are misleading. It would be more accurate to say that Antoinette was bullying you. When I read your article, my first thought is that for thousands of years Africans and lions shared the same land, and neither killed off the other. The hunting and habitat loss you describe are caused by Americans and Europeans."

Daisha looked despairingly at Baba Mutegi, "I can't write that can I?"

"Is that your perspective?"

"Well I understand her point. I agree with it, I guess. But it seems like it is not related to science."

"What would you say that it is related to?"

Daisha thought and responded uncertainly, "Maybe history? Or politics?"

After thinking about the exchange for a moment, Baba Mutegi responded to the group,

> "Science is a study of the natural world. It is done by humans. It is done for humans. So, we don't need to feel compelled to remove the human element. We don't have to clean it up or sanitize it. We can provide an authentic retelling—and even if that retelling delves into human error, human bias, exploitation, accident, racism, theft, jealousy, et cetera—it can still be an authentic treatment of science."

Mutegi's response was aimed at addressing a common notion of science, which is often misrepresented as being largely uninfluenced by the humans who conduct it. The popular misrepresentation accepts the influence of scientific genius, but rejects more unseemly human influences. Mutegi, perceiving this to be the source of Daisha's apprehension, worked to re-present science to the science writers as a two-sided endeavor: one that simultaneously included genius and idiocy, benevolence and exploitation, liberation and oppression. Throughout the three months that science writers worked to develop their articles, they pushed one another toward more authentic representations of science.

Theoretical Grounding

While the practice of reconceptualizing science content reflects each of the three theoretical perspectives (Codrington, 2014; Goduka, 2005; Mutegi, 2011), it is especially illustrative of Codrington's commitment to liberatory practice. Through her questions, Codrington reminds us that "systems of science [are] epistemologically and ontologically hegemonic in nature" and that systems of science "perpetuate cycles of racism, colonialism, and Western values steeped in dominance over people and consumption of natural resources."

Daisha's reticence about identifying European and American adventure seekers as one of the causes of lion extinction illustrates the extensiveness of the hegemony that Codrington identifies. So thorough is the hegemony that even 15-year-old Daisha questioned the epistemological validity of identifying White complicity in a sanctified science space. The epistemological and ontological hegemony position us to reject in toto the possibility that science can have any relationship with exploitation or oppression. In part, reconceptualizing science content requires that we explicitly identify and reject hegemonic ideas that serve as a barrier to meaningful science critique. Such critique of science practice includes (but is not limited to) critique of (a) the motives behind, (b) the methods of, and (c) the results of people's use and study of nature. Such critique also requires clear distinctions between those practices that are universally shared by all humans and those that are unique to one or more racial or cultural groups.

Reconceptualizing the Role of the Instructor

Reconceptualizing the role of the instructor requires a shift from a student–teacher relationship in which students and teachers work at cross-purposes toward a relationship characterized by joint productive activity.

Description of Practice

In the very first meeting of the science writers, Baba Mutegi asked each science writer to draft a one-page autobiography. In describing the exercise, he encouraged the science writers to write "what you think is important for the world to know about you." After composing an initial draft, the science writers and Baba Mutegi exchanged their autobiographies for peer review. During peer review, science writers were to identify for one another how they perceived the autobiographies from the perspective of a reader. While there are multiple layers of purpose in this exercise, three stand out as most salient. First, the exercise encourages the science writers to be self-reflective and to think through how each wants to present herself to a larger audience. Second, it provides an opportunity for science writers to identify consistency between what they want readers to perceive in their writing and what readers actually perceive in their writing. Third, it provides science writers an opportunity to build community by getting to know one another through the autobiographies and through the discussion generated by the autobiographies.

As science writers asked questions and shared impressions of one another's biographies, it became clear to Baba Mutegi that their responses were overly euphemistic. What is more, science writers worked to avoid providing any commentary, or suggestions that might be received as criticism. Baba Mutegi identified this as problematic and drew from an ethic of collective responsibility and an ethic of love to address it. "Antoinette, you did not identify any grammatical errors in Daisha's autobiography. Did she have any?"

"I guess."

"Is your answer, 'Yes' or 'No'?"

"Yes. She did have some errors."

"Did the errors make it difficult for you to grasp what she was trying to say about herself?"

"Hmmm. Maybe... Only once or twice."

"Why did you not identify them?"

"I don't know."

Addressing the group as a whole, Baba Mutegi pointed out that,

"Sometimes, people are uncomfortable criticizing one another. The discomfort comes from fear of being hurtful or being perceived as mean-spirited. For us, however, we have a responsibility to one another. Now Daisha is my sister... And when she leaves this room, or when she sends out her work, I want Daisha and her work to get the best reception possible. So, when she is in this room, it is my responsibility to do everything I can to help her improve herself and her work and to become the strongest writer she can be. I expect that she would do the same for me. So, Antoinette, we have to critique one another. Even though it may be uncomfortable, we also critique one another because it makes us all better and we do it from a spirit of love."

Kiona added, "I did the same thing Antoinette did. I think we just get used to not saying anything that could be perceived as mean. I don't want to be known as the mean girl. But, I understand and I really like this perspective. I like the idea of critiquing with a spirit of love."

Baba Mutegi asked the group, "So, if we think about critique as our responsibility to one another, what would you say about critique and our responsibility to the reader?"

Antoinette smiled, "I know where you're going with that question."

"Did I lead the question too much?"

Antoinette laughed, "Yes. So really it is the same thing. We critique each other so that we can be the best writers possible. But we also critique each other so that we can produce the best work possible."

Baba Mutegi responded, "Yes. Here too we do it in a spirit of love. We write books (and other materials) for Black children. And just like Daisha is our sister, the young children who read what we write, they are our little brothers and little sisters. We love them and we have a responsibility to give them our best."

The ethic of collective responsibility and the ethic of love were instrumental in shaping the organizational culture and the interactions of the science writers. Although it took repeated prompting, over time, science writers grew more comfortable sharing collective responsibility for the ideas and the work products generated.

Theoretical Grounding

Although the ethic of collective responsibility and the ethic of love are not explicitly addressed in either Codrington's (2014) or Mutegi's (2011) theoretical perspectives, these ethics are consistent with the overall approach and purpose presented by each author. Goduka's (2005) presentation of eZiko, however, is much more explicit in articulating not only these ethics, but also the positionality that teachers should have in relation to the students in their charge. Recall that the phrase *eZiko siPheka siSophula* is translated as "gathering around the hearth to cook and dish out." Recall also that the collective and communal characteristic of this metaphor is reflected in the positionality of the participants "we cook" (siPheka) and "we dish out" (siSophula). eZiko emphasizes for us the idea that teachers and students should be working collectively to nurture and provide sustenance to the community.

This approach can be contrast against the role teachers are expected to take in modern American schooling, especially the type of schooling to which African American students are subjected. Modern American schooling does not enjoin teachers and students together under an ethic of collective responsibility. Instead teachers are expected to take on a policing function, wherein they are charged with ensuring compliance with a broad range of expectations set at distant places. Neither does modern American schooling foster an ethic of love. Instead teachers are driven by the need to meet requirements mandated by a broad array of people, organizations, and institutions.

In a satirical essay on schooling, New York State Teacher of the Year John Taylor Gatto writes about the policing function that he enacted as a classroom teacher. He also alludes to the policing to which he was subjected as a teacher. Both forms of policing function to (a) eliminate the ethics of collective responsibility and love and (b) inculcate an adversarial relationship between teachers and students.

> *The second lesson I teach is class position.* I teach that students must stay in the class where they belong. I don't know who decides my kids belong there but that's not my business. The children are numbered so that if any get away they can be returned to the right class. Over the years the variety of ways children are numbered by schools has increased dramatically, until it is hard to see the human beings plainly under the weight of numbers they carry. Numbering children is a big and very profitable undertaking, though what the strategy is designed to accomplish is elusive. I don't even know why parents would, without a fight, allow it to be done to their kids.
> In any case, that's not my business. My job is to make them like being locked together with children who bear numbers like their own. Or at the least to

endure it like good sports. If I do my job well, the kids can't even *imagine* themselves somewhere else, because I've shown them how to envy and fear the better classes and how to have contempt for the dumb classes. Under this efficient discipline the class mostly polices itself into good marching order. That's the real lesson of any rigged competition like school. You come to know your place. (Gatto, 1992, pp. 4–5)

Discussion

In this chapter, we began with an assertion made by Amos Wilson that, "the problems that Black children face in schools… are political problems." To address this assertion, we (a) identified three theoretical approaches that speak to the political problems confronting Black children; (b) identified three ways that science education could be reconceptualized to draw from these theoretical approaches; and (c) provided examples of this reconceptualized science teaching and learning.

Notes on Systemic Implementation

A common question we receive in response to the reconceptualization described is, "How do we implement this modified vision of science teaching and learning?" By "implement" questioners are often referring to large-scale systemic implementation. The exasperation with which the question is typically asked suggests that the questioner does not believe that this modified vision could be implemented on a large scale. We agree.

The reconceptualization (of content, purpose, and relationship) that we describe in this chapter is necessary first because the current and prevailing conceptualization is detrimental to Black children, but also because it is so deeply and systemically entrenched into the fabric of Western culture. Our notion of STEM as an infallible way of knowing is so pervasive that to characterize a product as "scientifically" proven or an idea as a "scientific" fact trumps all further discourse. The idea that the purpose of schooling is job preparation is so pervasive that it is a truism. One of the primary reasons that we encourage children to stay in school is so that they can "get a good a job." The idea that a teacher's primary responsibility is to "control" a group of students pervades the expectations of everyone involved in education: administrators, teachers, students, and even parents.

These prevailing conceptualizations grow out of a host of deeply embedded social structures, cultural practices, histories, shared assumptions, policies, and practices. In addition to the structural impracticality of systemically implementing the suggested reconceptualization, there is the barrier of collective will. The majority of people do not see the prevailing conceptualizations as problematic. There are large numbers of people who have a great deal invested in the system as it is currently structured. Against the weight of deeply entrenched systemic structures and large numbers of people who are either committed to business as usual or indifferent to change, we agree with those questioners who might argue that systemic implementation is unlikely.

Notes on Local Implementation

We believe that it is more feasible to advocate for and to support local implementation. Here the efforts of individuals (such as parents, teachers, and paraprofessionals) and independently run organizations (such as home schools, private schools, charter schools, and civic and community organizations) have the potential to positively impact African American science learners. Individuals and independently run organizations have a greater ability to implement reconceptualized versions of science teaching and learning in their own educational space. What is more they can engage in local implementation in spite of and without changing the deeply entrenched systemic structures that make large-scale implementation impractical.

We will also point out that it is among the ranks of committed individuals and independently run organizations that we are more likely to find people who are not (a) committed to business as usual and (b) indifferent to change. Many of these individuals have already recognized and accepted Amos Wilson's assertion. It is this recognition and acceptance that has led them to seek alternative spaces (whether home school, private school, charter school, or the classroom) in which to effect change in the lives of Black children. Reconceptualizing science teaching and learning with these educators is less a matter of convincing them that it is necessary and more a matter of working together to determine how best to implement it.

Notes on Local Reeducation

For those individuals and independently run organizations interested in reconceptualizing science teaching and learning, we recommend three broad areas of re-education that will support such efforts. The first area of reeducation should position us to *understand the current educational approach as (a) systemically flawed, and (b) an instrumental tool in support of systemic racism.* Here we would argue that the disparate educational outcomes we see (especially the racially disparate outcomes) are not attributable to any isolated features of education as currently enacted. For example, while standardized tests might be problematic in many ways, the elimination of standardized tests will not radically improve the education of Black children. Standardized tests are one feature of an educational system that works to marginalize Black children. Tracking, high teacher turnover, the lack of Black teachers, irrelevance of the curriculum, and disciplinary practices are a few of the many other features of an educational system that works to marginalize Black children.

As a start to better understanding how the current approach is systemically flawed and a support for systemic racism, we recommend an exploration of the writing of critical curriculum theorists (Freire, 1985; Gatto, 1992; Illich, 1971; Kozol, 1991) and select educational historians (Chou, Lewis, & Watkins, 2001; Watkins, 2001; Woodson, 1933/2000).

The second area of reeducation should position us to *understand the nature of systemic racism as well as the various ways it has been established and is currently maintained.* We regard this understanding as essential in that it influences what we regard as fundamental educational problems as well as worthy educational goals. Is the cultural dissonance between White middle class female teachers and Black urban dwelling students a fundamental educational problem? When examining resources, how much should be committed to educating teachers about Black students? Would more be gained by committing those resources to efforts aimed at helping parents of Black students to be more effective advocates for their children?

As a start to better understanding the nature of systemic racism, we recommend a set of books (Welsing, 1995; Wilson, 1990; Wright, 1984; X & Haley, 1964) and documentaries (DuVernay, 2016; Hampton, 1986; Pollard, 2015; Tickell, 2007) that document and explain systemic racism both historically and in contemporary times.

The third area of reeducation should position us to *understand the fallibility of science both in its practice and in its implementation*. Reconceptualization of science content is often made difficult by the perception that science is sacrosanct. In order to support the efforts of individuals and independently run organizations interested in reconceptualizing science content, it might help to disabuse them of the notion that science is little more than an inviolate cannon of knowledge. Additionally, if done well, an exploration of more authentic representations of science would also provide practitioners with grist for critical explorations of science practice and implementation.

Fortunately, there is no shortage of resources for reconceptualizing science content. Each year journalists produce numerous social, cultural, and historical accounts on a wide range of science-related topics that expose the fallibility of science if not the impropriety in science practice and application. These accounts deal with food culture and consumption (Moss, 2013; Pollan, 2008; Schlosser, 2001), medical practice and malpractice (Angell, 2005; Blaser, 2014; Hornblum, 1998), human subject abuse (Jones, 1981/1993; Karlawish, 2011; Washington, 2006), the detrimental effects of technology (Carr, 2011; Mayer-Schönberger & Cukier, 2014; Postman, 1985), and environmental degradation (Carson, 2002; Egan, 2017; Perdew, 2017), among a host of others.

Note

1. The journals examined included: *Science and Children*, *Science Scope*, and *The Science Teacher*.

References

Allen, R. L. (2004). Whiteness and critical pedagogy. *Educational Philosophy and Theory, 36*, 121–136.

Angell, M. (2005). *The truth about drug companies: How they deceive us and what to do about it*. New York, NY: Random House.

Blaser, M. J. (2014). *Missing microbes: How the overuse of antibiotics is fueling our modern plagues*. New York City, NY: Henry Holt.

Carr, N. G. (2011). *The shallows: What the internet is doing to our brains*. New York, NY: W. W. Norton.

Carson, R. (2002). *Silent spring 40th anniversary ed.* Boston, MA: Houghton Mifflin.

Chou, V., Lewis, J. H., & Watkins, W. H. (2001). *Race and education: The roles of history and society in educating African American students*. Boston, MA: Allyn and Bacon.

Codrington, J. (2014). Sharpening the lens of culturally responsive science teaching: A call for liberatory education for oppressed student groups. *Cultural Studies of Science Education, 9*, 1015–1024.

Cohen, G. L., Garcia, J., Purdie-Vaughns, V., Apfel, N., & Brzustoski, P. (2009). Recursive processes in self-affirmation: Intervening to close the minority achievement gap. *Science, 324*, 400–403.

DuVernay, A. (Writer). (2016). 13th. In A. DuVernay, H. Barish & S. Averick (Producer).

Egan, D. (2017). *The death and life of the Great Lakes*. New York City, NY: W W Norton.

Freire, P. (1970). *Pedagogy of the oppressed*. New York, NY: Herder & Herder.

Freire, P. (1985). *The politics of education: Culture, power and liberation*. New York, NY: Bergin and Garvey Publishers, Inc.

Gatto, J. T. (1992). *Dumbing us down: The hidden curriculum of compulsory schooling*. Philadelphia, PA: New Society Publishers.

Goduka, N. (2005). Eziko: Sipheka sisophula. Nguni foundations for educating/ researchingfor sustainable development. *South African Journal of Higher Education, 19*, 467–481.

Goduka, N., Madolo, Y., Rozani, C., Notsi, L., & Talen, V. (2013). Creating spaces for eZiko Sipheka Sisophula theoretical framework for teaching and researching in higher education: A philosophical exposition. *Indilinga: African Journal of Indigenous Knowledge Systems, 12*(1), 1–12.

Hampton, H. (1986). *Eyes on the prize* [Film-Documentary]. Boston, MA: Blackside, Inc.

Hornblum, A. M. (1998). *Acres of skin: Human experiments at Holmesburg prison*. New York, NY: Routledge.

Illich, I. (1971). *Deschooling society*. New York, NY: Harper & Row.

Jones, J. H. (1981/1993). *Bad blood: The Tuskegee syphilis experiment*. New York, NY: The Free Press.

Karlawish, J. (2011). *Open wound: The tragic obsession of Dr. William Beaumont*. Ann Arbor, MI: University of Michigan Press.

Kozol, J. (1991). *Savage inequalities: Children in America's schools*. New York, NY: Harper Collins.

Lewis, B. F., Pitts, V. R., & Collins, A. C. (2002). A descriptive study of pre-service teachers' perceptions of African American students' ability to achieve in mathematics and science. *Negro Educational Review, 53*, 31–42.

Macedo, D. P. (1993). Literacy for stupidification: The pedagogy of big lies. *Harvard Educational Review, 63*, 183–206.

Mashoko, D. (2014). Indigenous knowledge for plant medicine: Inclusion into school science teaching and learning in Zimbabwe. *International Journal of English and Education, 3*, 528–540.

Mayer-Schönberger, V., & Cukier, K. (2014). *Big data: A revolution that will transform how we live, work, and think*. Boston, MA: Mariner Books.

Morton, C. H. (2017). Girls (STEM) Insitute (GSI): Changing the word by challenging, empowering and transforming girls and young women. *The Lighthouse Almanac, 1*(1), 8–10.

Morton, C. H., & Smith-Mutegi, D. (2017). Informal STEM learning: Impacting Black females' self efficacy and interest in STEM careers. In E. Galindo & J. Newton (Eds.), *Proceedings of the 39th annual meeting of the North American Chapter of the International Group for the Psychology of Mathematics Education* (pp. 1129). Indianapolis, IN: Hoosier Association of Mathematics Teacher Educators.

Moss, M. (2013). *Salt sugar fat: How the food giants hooked us*. New York, NY: Random House.

Muller, P. A., Stage, F. K., & Kinzie, J. (2001). Science achievement growth trajectories: Understanding factors related to gender and racial-ethnic differences in precollege science achievement. *American Educational Research Journal, 38*, 981–1012.

Mutegi, J. W. (2011). The inadequacies of "science for all" and the necessity and nature of a socially transformative curriculum approach for African American science education. *Journal of Research in Science Teaching, 48*, 301–316.

Mutegi, J. W. (2015a). *Kayla's first chickens: Book four*. Indianapolis, IN: Sankore Institute.

Mutegi, J. W. (2015b). *Tale of the vanishing buffalo*. Indianapolis, IN: Sankore Institute.

Mutegi, J. W. (2016). *Ronnie's great idea*. Indianapolis, IN: Sankore Institute.

Mutegi, J. W., Lewis, N., & Smith-Mutegi, D. (2017). *Socially transformative science curriculum: Encouraging critical perspectives among preservice teachers*. Paper presented at the annual meeting of the American Educational Research Association, San Antonio, TX.

Mutegi, J. W., & Morton, C. H. (2012). Socially transformative science pedagogy for African American males: Dispatches from the Vanguard. *African American Learners, 1*(2).

Myers, C. (2014, March 16). The apartheid of children's literature, *New York Times, 163*(56442), SR1.

Myers, W. D., & Myers, C. (2014, March 16). When are the people of color in children's books? *New York Times, 163*(56442), 1.

Norman, O., Ault, C. R., Bentz, B., & Meskimen, L. (2001). The black-white "achievement gap" as a perennial challenge of urban science education: A sociocultural and historical overview with implications for research and practice. *Journal of Research in Science Teaching, 38,* 1101–1114.

Perdew, L. (2017). *The great pacific garbage patch.* Edina, MN: Abdo Pub.

Pitts Bannister, V. R., Davis, J., Mutegi, J. W., Thompson, L. R., & Lewis, D. D. (2017). "Returning to the root" of the problem: Improving the social condition of African Americans through science and mathematics education. *Catalyst: A Social Justice Forum, 7*(1), 4–14.

Pollan, M. (2008). *In defense of food: An eater's manifesto.* New York, NY: Penguin Press.

Pollard, S. (Producer). (2015). *Slavery by another name.* San Francisco, CA.

Postman, N. (1985). *Amusing ourselves to death: Public discourse in the age of show business.* New York, NY: Viking.

Riggs, M. (Writer). (1987). *Ethnic notions* [DVD]. In M. Riggs (Producer). San Francisco, CA: California Newsreel.

Schlosser, E. (2001). *Fast food nation: The dark side of the all-American meal.* Boston, MA: Houghton Mifflin.

Taleni, V., & Goduka, N. (2013). *Perceptions and use of indigenous leafy vegetables (ILVs) for nutritional value: A case study in Mantusini Community, Eastern Cape Province, South Africa.* Paper presented at the International Conference of Food and Agricultural Sciences (IPCBEE), Singapore.

Tickell, P. (Writer). (2007). *Racism: A history* [Videorecording]. In D. Okuefuna (Producer). Hamilton, NJ.

Washington, H. A. (2006). *Medical apartheid: The dark history of medical experimentation on Black Americans from Colonial Times to the present.* New York, NY: Doubleday.

Watkins, W. H. (2001). *The White architects of Black education: Ideology and power in America, 1865–1954.* New York, NY: Teachers College Press.

Welsing, F. C. (1995). *The Isis papers: The keys to the colors.* Chicago, IL: Third World Press.

Wilson, A. N. (1990). *Black-on-Black violence: The psychodynamics of Black self-annihilation in service of white domination.* New York, NY: African World Infosystems

Woodson, C. G. (1933/2000). *The mis-education of the Negro.* Chicago, IL: African American Images.

Wright, B. E. (1984). *The psychopathic racial personality and other essays.* Chicago, IL: Third World Press.

X, M., & Haley, A. (1964). *The autobiography of Malcolm X.* New York, NY: Ballantine Books.

Broadening Millennials' Participation in STEM and the Teaching Professions Through Culturally Relevant, Place-Based, Informal Science Internships

JACQUELINE LEONARD, SCOTT A. CHAMBERLIN, ELSA BAILEY, GEETA VERMA, AND HELEN DOUGLASS

Abstract

The empirical work reported in this chapter was a part of a larger study. The chapter accomplishes two primary goals. First, we present a model that broadens interns' knowledge of and participation in culturally relevant, place-based teaching through engagement with underrepresented elementary-aged children in informal science programs. Second, we examine the efficacy of this model to enable interns to engage younger children in science and broaden their own participation in STEM or teaching careers. Twenty-five interns participated in Saturday science academies and/or summer science camps. Results indicated the interns' training reinforced content knowledge, provided guidance to teach science lessons, and allowed for reflection on professional goals. While the interns who participated in the summer camps had stronger content knowledge than the interns who participated in Saturday academies, the interns in the Saturday academies reported more robust opportunities to teach compared to interns in the summer camp. Nevertheless, the program succeeded in maintaining or promoting excitement about STEM and teaching professions. We conclude that the program was moderately successful in providing interns with opportunities to learn and teach science.

Introduction

The purpose of this chapter is to share our research model for recruiting and training high school and college-aged students to work as interns in informal science programs that targeted urban children and youth. We conducted a two-year study that used culture and place as hooks to broaden science content knowledge and participation in science, technology, engineering, or mathematics (STEM) fields. The goals of the larger study were to (a) provide underrepresented children and youth (ages 8–12) with opportunities to learn about paleontology and natural science and to give them exposure to STEM careers through interactions with STEM interns and professionals; and (b) to recruit and train interns to enhance their science content knowledge to reach and engage other youth. The current study focuses on the second goal to recruit and train interns from underrepresented backgrounds for the broader purpose of recruiting them into careers in STEM or education. We examined the instructional practices of 25 interns who helped to facilitate learning about paleontology during Saturday science academies, and natural science topics during a summer science camp, as well as their interest in STEM or teaching as a profession.

Teaching, once considered a "stable, high-status profession" (Ladson-Billings, 2005, p. 2), continues to experience shortages in the subject areas of mathematics, science, and special education (Maranto & Shuls, 2012). According to the U.S. Department of Education (2013), 46 states had science teacher shortages during the 2009–2010 academic year. Recruiting and retaining high-quality teachers with expertise in science and mathematics to work in hard-to-staff schools is particularly challenging. These challenges are further exacerbated when it comes to recruiting teachers of color and teachers to work in diverse communities (Evans & Leonard, 2013; Waddell & Ukpokodu, 2012).

According to Vilson (2015), the teaching profession remains predominantly white (82%), creating what has been characterized as a diversity gap (Hrabowski & Sanders, 2015; Waddell & Ukpokodu, 2012). This *demographic gap* between the teacher workforce and K–12 students has negative effects upon the achievement variables of students of color because of cultural and linguistic disconnects in the teaching-learning process (Ladson-Billings, 2009; Lucas & Villegas, 2011; Sleeter & Thao, 2007). Teachers of color are more likely to stay in high-need schools and work with children of color longer than white teachers (Hrabowski & Sanders, 2015). Racial diversity among school staff also decreases the likelihood of teacher mobility (Moore, 2012).

To prepare diverse teachers with caring dispositions (Cochran-Smith, 2004) and STEM backgrounds, it is necessary to provide those who have not considered STEM or teaching careers with opportunities to teach (Martin-Hansen, Puvirajah, & Verma, 2012). Informal learning environments provide future teachers with opportunities to teach and learn from their practices while engaging children in authentic science and mathematics experiences (Leonard, Chamberlin, Johnson, & Verma, 2016). Opportunities to engage in authentic science and mathematics experiences are necessary to positively change the trajectory of underrepresented students' participation in STEM fields. One of the best predictors of STEM career interest after high school is youth's interest in STEM before high school, which suggests the importance of early intervention (Sadler, Sonnert, Hazari, & Tai, 2012). It is important to not only examine the interns' STEM interest and experiences as a recruitment tool but also their influence on youth participants' STEM engagement as a result of their practices. Thus, this chapter focuses on STEM interns' (a) opportunities to learn and teach science; (b) ability to engage children and youth in STEM; and (c) interest in STEM fields or teaching.

Theoretical Framework

The theoretical framework that supports the current study is critical race theory (CRT) and the related constructs are culturally relevant pedagogy (CRP) and place-based education (PBE). Critical race theory in education draws from the traditions of Critical Legal Studies (CLS) and radical feminism (Delgado & Stefancic, 2012). Proponents of CRT understand the importance of: (a) acknowledging the permanence of race and racism in the U.S.; (b) challenging claims of meritocracy and colorblindness; (c) teaching for social justice; (d) validating the experiences and knowledge of people of color; and (e) supporting counterstories to unmask narratives that subjugate others (Bell, 1980). However, CRT asserts that the impetus for change must come from the voices, narratives, and counterstories of marginalized individuals. CRT is an appropriate framework to study STEM learning and engagement, particularly from the perspectives of interns from underrepresented backgrounds, because it provides them with opportunities to share their experiences and counterstories (Solorzano & Yosso, 2002). Furthermore, CRT provides a framework to examine interns' teaching in culturally relevant and place-based, informal learning environments.

CRP is a theory developed by Ladson-Billings (1995) to support African-American children's learning through academic success, cultural competence, and critical consciousness. Rather than adhering to deficit theory, teachers and researchers should view students' culture as an asset upon which to build new knowledge. In educational settings, it is critical to understand other ways of knowing and learning STEM (Greer, Mukhopadhyay, Powell, & Nelson-Barber, 2009). Culture is embedded in the natural world and influences every aspect of our lives. CRP provides underrepresented students with an opportunity to learn in a *third space*, which is defined as "newly recontexualized content and an environment that surrounds learning that content" (Lipka, Yanez, Andrew-Ihrke, & Adam, 2009, p. 266), where ethnic ways of knowing are valued alongside dominant canons of knowledge (Gay, 2010). CRP is an attempt to use the cultural capital that students bring from home and community as a springboard for learning. Cultural capital embodies the norms, ideologies, language, behavior, mores, and practices of a particular group and is transmitted to children as cultural knowledge (Bourdieu, 1973; Howard, 2003). Using interns from the same backgrounds as the students not only draws upon cultural capital but also supports the development of cultural competency, which is learning about one's own culture as well as another culture (Ladson-Billings, 1995, 2009). In the current study, the culture of the community was African American and Mexican American.

In addition to CRP, a construct that is experiencing resurgence as a supplement to traditional instruction is PBE (Howley *et al.*, 2011). Dewey (1916), an early proponent of PBE, asserted that experiential learning allowed children to benefit from their natural surroundings. PBE operates under the assumption that education is not place-less, but grounded somewhere in order for learning to occur (Aikenhead, Calabrese Barton, & Chinn, 2006). Place-based learning has roots in the local community—its culture, economy, history, and the arts (Gruenewald, 2003; Monk *et al.*, 2014). Drawing upon community resources, PBE allows teachers and students to tap the knowledge of scientific experts and community leaders to enrich STEM learning and make connections to the real world. In this study, we used place to immerse the interns in local habitats and natural environments to strengthen their science content knowledge and attract them to the fields of STEM or teaching. The interns, who are the focus of this chapter, learned about science and how their lives fit into the spaces they occupied, which had the potential to reinforce belonging in STEM and education communities (hooks, 2009).

Background of the Study

In order to ground this study in the literature, it is important to understand different types of internships and the value of teaching in informal science learning environments.

Field-Based Experiences

Field experiences have been defined "as a variety of early and systematic P–12 classroom-based opportunities in which teacher candidates (TCs) may observe, assist, tutor, instruct, and/or conduct research" (Capraro, Capraro, & Helfeldt, 2010, p. 131). However, not all field experiences are created equal, and Capraro *et al.* acknowledge that field experiences may take place in multiple settings including informal learning environments. Leonard, Barnes-Johnson, Dantley, and Kimber (2011) provided elementary TCs with an additional field experience beyond practicum and student teaching to work in community-based settings. In this context, TCs were paid stipends to lead instruction in afterschool and Saturday academies in Pennsylvania. Urban students learned earth and space science by engaging in inquiry-based, hands-on activities that required collecting data on student-made gliders, veggie shuttles, and Alka-Seltzer rockets (Leonard et al., 2011; Leonard & Oakley, 2006). While teaching practices are not easily changed, the TCs were able to hone their practice by working with small groups of students, and the value of hands-on activities was reinforced. Classroom observations revealed urban students were highly motivated, eager to learn, and enjoyed working with the interns, several of whom came from similar racial and ethnic backgrounds.

Borgerding (2015) conducted a study that used summer internships to recruit and attract STEM majors, who had not previously considered teaching as a career. Results of a prior study revealed that internships were not enough to attract interns into education (Worsham, Friedrichsen, Soucie, Barnett, & Akiba, 2014). Thus, making a difference or working with underrepresented students could be an impetus for millennials. The sample included five Noyce scholars who worked as interns in a residential Upward Bound program that focused on teaching mathematics and science to urban high school students for three to four weeks. A qualitative approach was used to examine how summer internships influenced the interns' career paths and ideas about teaching and learning science. Case studies revealed that two interns developed high interest in teaching, two developed marginal interest, and one developed reduced interest in teaching. These findings revealed the importance of informal science placements in recruiting and training STEM majors to become teachers, which is critical to the Robert Noyce Teacher Scholarship program and other teacher credential programs (Bishoff, French, & Schaumloffel, 2014; Capraro et al., 2010). In the current study, we recruited interns with STEM and education backgrounds to participate in the study to examine their learning and teaching practices, influence on the STEM engagement of children and youth, and interest in STEM or education.

Informal Science Education

Informal science education can take place in a variety of contexts (Verma, Puvirajah, & Webb, 2015). Science camps and afterschool programs provide opportunities for young children to develop "cognitive abilities to engage in STEM content and problem-solving activities" (DeJarnette, 2012, p. 80). Moreover, informal science programs have the potential to foster inquiry, science literacy, and a sense of place and belonging among young children (hooks, 2009; Rye et al., 2012). In such settings, learners acquire science processes by manipulating one variable to learn how it affects the results (Rye et al., 2012).

Wallace and Brooks (2015) used an experiential, informal science context to provide opportunities for TCs to work in a K–7 summer camp. Providing teaching opportunities in informal learning environments also allowed TCs to engage in inquiry-based teaching and to develop a science identity (Carlone & Johnson, 2007). The researchers contended that learning to teach science in an informal learning environment "had strong potential to influence elementary TCs' enjoyment of teaching science and understandings about how inquiry-based strategies develop conceptual knowledge" (Wallace & Brooks, 2015, p. 178). Ethnographic methods were used to collect data and conduct case studies on 12 TCs. Findings revealed that TCs developed autonomy, peer collaboration, and strong ties with children, which strengthened self-efficacy.

In a study known as EnvironMentors (Monk et al., 2014), researchers partnered with an existing GEAR UP program at a local high school. The goals of the study were to (a) conceptualize and conduct an environmental science themed research project with a mentor; (b) increase environmental literacy and awareness of environmental issues facing Louisiana and the Gulf Coast; and (c) increase college awareness by exposing students to a university environment. Results of the program revealed three important findings: (a) students succeeded in the program, performed well on high school curricula, and enrolled in college; (b) students enjoyed the program and showed increased interest in environmental science at the postsecondary level; and (c) mentors found the program to be rewarding, and they enjoyed sharing their knowledge and expertise with students. While the mentors in the EnvironMentors study were scientists and not interns, the findings informed the current study as it related to developing interns' interest in STEM careers.

In another study, Akom (2011) linked community health issues to environmental education as it pertained to the presence of lead and asbestos in housing units and the presence of polychlorinated biphenyls in poor communities. Akom (2011) encouraged youth to study links to race, space, place, and waste by using the Geographic Information System and ground trooping (i.e., work at the grassroots level) to collect data on biohazards to increase community awareness and to advocate for environmental justice. Thus, linking informal science education to environmental justice can motivate students to learn and provide them with lifelong skills. In the study reported here, we used curriculum that made links to the environment, natural resources, and ecojustice to motivate interns to consider STEM or teaching careers.

Research Questions

Three research questions guided this study on 25 interns' learning and instruction, ability to engage underrepresented students in science, and interest in majoring in STEM or education.

1. How did the interns' training and field experience (i.e., internship) prepare them to teach science during Saturday academies and/or summer camp?
2. How well did the interns engage children and youth in science activities during Saturday academies and/or summer camp? Was cultural relevance or place evident?
3. How did the internship influence the interns' interest in STEM fields or teaching?

The Context of the Study

Researchers at two universities in the western U.S. collaborated during the study presented in this chapter. Two scientists partnered with the researchers to develop and implement the science curriculum. A retired geologist implemented curriculum on paleontology during the Year 1 study, and a soil scientist employed by the U.S. Department of Agriculture implemented lessons on the environment to learn about natural science in Year 2. The interns were a core part of both study years. The process for recruiting the interns included a review of letters of interest and one-page résumés. Interns with a background in science, mathematics, or education (e.g., college major or expressed college major) were given priority for hiring.

Prior to working with children, the interns completed four hours of training led by project staff to become familiar with logistics, science content, and the objectives of each lesson. In Year 1, interns attended sessions led by the geologist to learn about the characteristics of dinosaurs and two different periods when they roamed the Earth (i.e., Jurassic and Cretaceous). In Year 2, interns had a review of the lessons they would teach related to environment and natural science and were exposed to the microscopes students would use to examine plant and animal cells to reinforce the importance of caring for the Earth.

The interns were paid a stipend to facilitate student learning in their group as the scientist or project staff presented key components of the lessons. They also assisted students as they worked in pairs or triads to complete the tasks. In most cases, children were assigned to interns based on gender and ethnicity. For example, girls were assigned to female interns and Spanish-speaking children, who were still developing English fluency, were assigned to Latinx[1] interns. We believed that cultural relevance and cultural competence could be developed within these groups and that stronger ties would develop between the interns and the children if they shared similar characteristics (Hrabowski & Sanders, 2015). We also encouraged children and interns from different ethnic backgrounds to mingle during snacks, lunch, free time, and field trips.

The lessons were conducted at local churches in fall 2012 and at a community-based center in summer 2013. The fall academies took place on four consecutive Saturdays where students had outdoor field experiences every other week at Dinosaur Ridge. The geologist led the field experiences, which allowed students to see actual fossils and measure dinosaur tracks. A soil scientist led composting and soil science activities, and the researchers led geology (e.g., layers of the Earth) and biology (e.g., cells) activities. In addition to these activities, students experienced a number of field trips to reinforce their learning of science content. Each intern worked with assigned students in both settings, interacting with them as they engaged in STEM activities. They also chaperoned students during field trips.

The Role of the Researchers

The researchers included two mathematics educators (principal investigator and senior personnel) at a land-grant university and a science educator (co-principal investigator) at an urban university. Both institutions were located in the Rocky Mountain West. The researchers were diverse in terms of race and ethnicity (one African American/Black; one Asian; and one European American). Their role during the study was to handle the logistics in terms of recruiting the interns and students, securing the facilities, organizing the curriculum and field trips, purchasing the materials and making them available, supervising the interns' work with children and youth, chaperoning field trips, and collecting data.

Methodology

We used qualitative methods to collect and analyze data in the study by employing thick descriptions of informal learning environments to understand the teaching-learning contexts (Creswell, 1998). Specifically, we examined the statements of interns and students, which are classified as counternarratives of their experiences in science and their engagement in STEM learning. Counternarratives can be used as a method (Solorzano & Yasso, 2002) to examine the experiences of those who have been marginalized in traditional science education. We also used the constant-comparison method (Strauss & Corbin, 1990) and open coding (Lewis & Maas, 2007) to examine the interns' counternarratives for emergent themes and patterns related to teaching practices, STEM engagement, and interest in STEM or education careers.

Data Collection

Data were collected across two years in the current study. The researchers, an external evaluator, and a doctoral student engaged in data collection. The researchers collected observational data in the form of field notes and photographs during both study years. They also administered the STEM Semantics Survey (Knezek & Christensen, 2008) in Year 1 (see Appendix A). The survey uses a 7-point Likert scale to assess attitudes toward STEM and STEM careers. The evaluator administered an open-ended questionnaire, which was developed and distributed to the interns to complete on the final day to gather data on their backgrounds, interests, and experiences during the internship (see Appendix B for questionnaire). For example, one question asked, "What stands out for you from your work this week?" To maintain anonymity, identification numbers were used to collect multiple sources of data on each intern. The survey and questionnaire allowed for the collection of demographic data as well (i.e., race/ethnicity and gender). The evaluator also conducted focus group interviews with student participants. The doctoral student conducted focus group interviews with interns during the study to gather data on their engagement with students.

Data Sources

Qualitative data sources included field notes, photographs, open-ended questionnaires, and transcripts of focus group interviews. Field notes were recorded by the researchers to maintain archival records on logistics and training sessions for the interns, lesson plans, and field trips. In a few cases, lessons were transcribed by project staff to capture teachable moments.

Photographs were taken by the research team to document science teaching and engagement. Focus group interviews were transcribed verbatim to gain additional information about the interns' and children's experiences. Lastly, the questionnaire completed at the end of the study provided insight into the interns' experiences during the internship itself.

Data Analysis

Data analysis consisted of open coding to find key words and terms in the data to describe the interns' teaching experiences or engagement (i.e., verbal or physical interactions) with students. These data were used along with questionnaires to find patterns in the data. Similar terms were then grouped into categories to find emergent themes. We also analyzed data collected from surveys, questionnaires, and focus group interviews to gauge interns' interest in a career in STEM or teaching.

The Sample

Sixteen interns participated in the Year 1 study in fall 2012. Eight interns (four males; four females; seven Mexican Americans, and one European American) worked at an academy that was located at a local church in a Mexican-American community. Eight interns (four males; four females; seven African Americans; one Biracial) worked as interns at an academy that took place in a local church in an African-American community. Twelve interns worked during the Year 2 summer camp that was hosted at a community-based site. Nine were males and three were females. Eight were African American and four were Mexican American. These interns facilitated learning with a total of 67 students (33 in Year 1 and 34 in Year 2). However, three of the 12 interns also worked during the Saturday academies (one African American; two

Mexican Americans). Thus, while the headcount by setting is 28, only 25 interns participated in the study. The students taught by the interns identified as African American (38.8%), Mexican American (50.7%), and two or more races (9.0%). Table 4.1 shows the demographics of the cohorts of interns as well as their educational level and college majors.

Table 4.1: Demographic Data for Summer Interns.

Demographics	Cohort 1	Cohort 2
Race/Ethnicity		
African American/Black	7	8*
Mexican American	7	4**
European American	1	0
Two or more races	1	0
Gender		
Female	8	3*
Male	8	9**
Academic Level		
High School	11	8**
College	5	4*
Majors		
STEM	3	3*
Education	2	1
Total Interns	16	12

*Includes 1 continuing student
**Includes 2 continuing students

Limitations of the Study

The results of this study are limited to the participants and settings where it took place and should not be generalized beyond this context to other underrepresented students. Moreover, given the competitive nature of the selection process, it is also possible that some interns may have reported higher levels of interest at the outset to be selected to participate in the study. Thus, we acknowledge that self-reports are less reliable forms of data than most other types of data collection, especially when affective ratings are considered (Anderson & Bourke, 2000).

Results

To answer the research questions, we relied on field notes, photographs, questionnaires, and focus group interviews to describe the interactions and learning that took place during the study. First, we describe lessons taught in each context to understand the content knowledge the interns needed to know in each setting. Second, counternarratives obtained from questionnaire responses and focus group transcripts were used to describe some of the teaching episodes. These excerpts were also analyzed by researchers to show emergent themes that arose in each of the teaching-learning contexts. Finally, questionnaires and focus group transcripts were analyzed to report the interns' interest in a STEM or teaching career.

Saturday Academies

Recall that 16 interns worked during Saturday science academies. The academies ran four consecutive weeks from 10 am to 12:30 pm (with a break for snacks). Half of the interns worked at a local church with 9 African-American children in October 2012, and other half worked at a neighborhood church with 22 Mexican-American children in November 2012. Because of the lower enrollment in the African-American community, the student–intern ratio was roughly 1:1 while it was about 3:1 in the Mexican-American community. This schedule allowed the geologists and researchers to work with both sets of students.

Paleontology Lessons. The interns attended a four-hour training session taught by the geologist to prepare them for the lessons they facilitated during the Saturday academies. They not only reviewed all of the lesson plans but also learned the meaning of various terms used in paleontology. For example, *bronto* means *thunder* and *apato* means *deceptive*. *Dactal* means *finger* and *raptor* means *thief*. The first lesson involved putting a dinosaur puzzle together and making a scaled drawing of a Brontosaurus, which interns learned was misnamed. The correct name is Apatosaurus. The second lesson was field-based as the interns chaperoned students to Dinosaur Ridge where they explored the museum and dinosaur fossils. In lesson three, interns helped students at the community-based center to create a timeline showing when dinosaurs lived on Earth in comparison to humans. The interns also encouraged students to complete a scavenger hunt that matched animals to their habitats. The final lesson took place at Dinosaur Ridge. Interns assisted students as they investigated fossilized dinosaur tracks (see Appendix C). For example, the Apatosaurus moved at a rate of 12–19 miles per hour. Students determined whether an Iguanodon was walking, trotting, or running by measuring the distance between the tracks (Leonard et al., 2016).

Interns' Learning and Preparation for Teaching. In terms of their own learning, interns reported on the questionnaire that they learned new information about dinosaurs as a result of their participation in the training:

> I now understand what is actually categorized as a dinosaur.
> Now I know the differences between types of dinosaurs.
> I now understand the differences between meat eaters, plant eaters, and animals that are not considered dinosaurs.

A focus group interview with three target interns not only revealed deep learning for the interns but also shed light on their science backgrounds:

> [The lesson] blew me away last week. I realize that I didn't know what the definition of a dinosaur was until then. There was pictures of a dinosaurs in water, they wasn't even dinosaurs. They were like...I would interpret as a dinosaur, but they weren't, and I like got an understanding of their size. When I think of a dinosaur, I think of all of them as big, but there's some that are smaller. And another amazing thing is the kids like when they first came, they were asked how would you know a dinosaur was a meat eater or plant eater. They knew that. They were like well, look at the claws that make them a meat eater. I wouldn't have known those clues would suggest that.

The same intern who made the above comment remarked:

> When I was a kid I didn't have any science. Like this year, this year is like one of my second years ever having any kind of science base learning.

A second intern in the focus group had a similar experience:

> I never had [science] growing up. Like some dinosaurs I thought was dinosaurs aren't dinosaurs. That's what I was taught when I was younger so come to find out the evidence [does not support it]. It's amazing!

These comments reveal shortcomings in large, public school systems that often focus on reading and mathematics while science is on the backburner (Leonard et al., 2011). These counternarratives reinforce the need to develop scientific knowledge among underrepresented students before they reach high school (Sadler et al., 2012).

Data obtained from questionnaires also suggest that learning was not restricted to the training sessions. Several interns commented that they learned while engaging students in the activities:

> We actually got to learn while teaching. I did not know as much as I do now about dinosaurs, and I thought it would be fun to learn about them while working with the youth. What stood out the most is that I was actually able to help out my younger peers and also get help from them.
>
> You help kids learn about dinosaurs, and you also learn about them, too.

Comments about learning to teach while teaching also occurred during focus group interviews with interns:

> It's easy teaching them, and they teach us, too, just as much as we teach them.

Three interns who also participated in a focus group described how lesson planning and organization of the materials were helpful in facilitating the lessons:

> ...everything [went] according to plan. There's programs where setbacks happen and things don't happen like it was suppose to happen. Everything that happened like when we went up to Dinosaur Ridge like the sandbox, everything was prepared.
>
> I like how everything is coordinated. Cause usually things is like, it's not planned out, and I just like how...we are... on schedule and all that.

> And just to add on to what they said, just how organized everything is you know. We have the lesson plan for the day so you know what you're doing that day and what all everybody needs to do.

These data suggest the interns were confident in doing their work because expectations were clearly articulated, and they knew what was coming next in the lesson plans.

Interns' Interactions With Students. Field notes, photographs, and student data were used to document the interns' involvement in the science lessons. Excerpts from student focus groups substantiate this claim:

> I guess I thought it was cool how we all got to on the first day, we all worked together and we made a puzzle and we made a big T-rex head. I thought it was cool how we were all working together. We didn't have any individual work. It was all hands-on activities.
> Building a dinosaur on the wall. Because we had to draw it first, and then we had to put it together.
> ...our first time, when we first met [the geologist] we had a puzzle and that was fun, working together, trying to get the big pieces together.

Another student emphasized learning the meaning of dinosaur names:

> On one of the days, I forgot how many weeks ago, we found the meaning of what the names mean. One of the dinosaur's name says "robber" and another word saying that it's not fearful. I made a saying, "I'm not afraid, and I got all the money."

This lesson was place-based because dinosaurs were found in the western region of the U.S., positioning dinosaurs and humans from a scientific and historical perspective.

In addition to activities at the community-based centers, interns were also highly engaged with students during the outdoor field activities. This was especially evident at Dinosaur Ridge as interns explained on their questionnaires how students demonstrated mathematics skills during a measurement activity:

> I noticed that the four girls I got to work with liked using the tape measurer to measure the dinosaurs and the calculators.
> Kids were working with mathematics, measurement, years, and fossils. They had to work hard and think so hard to find the answers.

Kids in my group were good at doing math and predicting what bones were for what dinosaur.

The kids were putting clues together to find out measurements an easier way. For example, the hip length...they would multiply the foot size by a number to get the hip length rather than measure the hip.

Emergent Themes. Using the constant-comparison method and open coding (Strauss & Corbin, 1990), the themes that emerged during the Saturday academies were *discovery, collaborative learning,* and *cohesive lesson plans.* It is evident from the data above that interns learned and made discoveries about dinosaurs during the Saturday internship. Not only did they discover what was and was not a real dinosaur, they also engaged in collaborative learning alongside the students (Vygotsky, 1978). This opportunity to engage in mutual learning is consistent with CRP as interns and student participants developed cultural competencies in African-American or Mexican-American ways of knowing and learning and crossed borders into the culture of science (Aikenhead, 1996; Ladson-Billings, 2009). For example, interns and students understood the power of names and naming in paleontology. The final theme was cohesive lesson planning. Although the lessons took place a week apart, the lesson plans were cohesive and helped to refine teaching and learning about paleontology. The interns found it easy to teach in this context as lesson plans built upon what was previously learned the week before.

Interest in STEM Field or Teaching. The results of the pre–post survey, as shown in Table 4.2, show no significant differences in science, mathematics, or engineering attitudes or attitude toward STEM careers. The only significant increase among this cohort of interns was attitude toward technology. This change is encouraging given the importance of technological advances in STEM, education, and society in general.

Table 4.2: STEM Semantics Survey ($n=15$).

	Science	Math	Engineering	Technology	STEM Career
Pre-test	5.7	4.8	4.8	5.9	5.6
Post-test	5.8	4.9	4.9	6.7*	5.9

*$p < 0.5$

We also analyzed the interns' responses on the open-ended questionnaire regarding their career interest. Six interns were interested in STEM/STEM-related fields (one science; one mathematics; one computer science; one technology; one animal science; and one pharmacy). Three implied that they had an interest in teaching after having participated in the internship. Thus, 56% of the interns who participated in the Saturday academies were interested in STEM/STEM-related careers or education after participating in the internship. Given that there were two STEM majors and two education majors (25%) at the outset, the number of interns interested in STEM after the study was a positive outcome. Excerpts obtained from the interns' responses to the questionnaire support this finding.

> *The study of geology interested me and the dinosaurs played a big part as well. I grew up loving dinosaurs and to this day I still do, so I thought this would increase my knowledge and experience with teaching kids.*
> *I wanted to gain more experience working with and teaching kids and loved that it focused on math and science.*
> *I enjoy working with children and I am interested in helping young people learn about science.*
> *It gives me experience with kids, and I feel that the kids liked me and may make me want to work with kids in my future.*
> *It enhanced my desire to become a teacher and experience this on a daily basis.*

The Summer Camp

Recall that 12 interns worked during a one-week summer camp at a community-based center in the western U.S. The one-week camp ran from 9 am to 4 pm Monday through Friday in July 2013 (with breaks for snacks and a one-hour lunch with free time). These interns worked with 34 children; the student–intern ratio was approximately 3:1. Specific activities consisted of: (a) observing the germination and root system of a Lima bean; (b) planting flowers in a community garden; (c) examining onion skin and cheek cells; (d) learning soil sampling techniques; (e) learning about the layers of the earth; and (f) making and recording observations of a compost pile. The compost pile was a weeklong activity that allowed the students to see how to start a compost pile as well as how it progressed into something that could be used to safely recycle ecologically friendly items. Children also went on field trips to the local Museum of Nature and Science, the local Botanic Garden, and Rocky Mountain National Park. Descriptions of the lessons and the field trips are given below.

Soil and Natural Science Lessons

During the summer camp, the majority of the lessons focused on soil science and natural science. Soil science is concerned with the formation, nature, ecology, and classification of soil, and natural science is concerned with the laws of nature and the physical world, including biology, chemistry, and physics. Two classroom-based lessons resonated with the interns and student participants: microscope lesson and soil texture lesson. The purpose of the microscope lesson was to help children learn the importance of microorganisms in the natural environment. This lesson was culturally relevant because students used their own cheek cells during the lab as well as onions. The purpose of the soil texture lesson was to build an understanding about the importance of soil as it related to: (a) habitat for animals and microorganisms; (b) medium for growing all types of plants; and (c) stable foundation for buildings and roads (U.S. Department of Agriculture, 2005). This lesson was also culturally relevant because it allowed students to develop critical consciousness on ecojustice while they learned about soil and composting (Akom, 2011; Ladson-Billings, 2009). It was also place-based because the red clay dirt was unique to the Rocky Mountain West.

The Microscope Lesson. The soil scientist, one of the mathematics educators, and the science educator led the students through the microscope activities. There were 10 microscopes along with tools and supplies to complete the tasks. Ten stations were set up around the multipurpose room for interns to work with two or three children on the activities. The first activity consisted of viewing ready-made slides of insects and one-cell microorganisms that are often found in pond water. Second, with the help of interns, the children learned how to prepare their own slides. To learn more about cells, student participants prepared onionskin slides. Working in small groups of two or three, the children first peeled a layer of skin from an onion slice and created a wet mount. Using protective gloves, the interns added a drop of iodine on top of the onionskin. Then the children examined the onion cells under a 400x microscope and looked for the cell nucleus that was stained with the iodine. They recorded observations and took notes in their journals. On day two, the children prepared a second slide to examine cheek cells. They first gently scraped the inside of their cheeks using a toothpick to create a wet mount. Then they mixed the cheek cells on the toothpick with a solution of blue dye and rubbing alcohol. Finally, students examined the slides under the microscope, made observations, and compared the cheek cells to the onion cells.

Soil Texture Lesson. The soil scientist led small groups of students through a hands-on method and a mechanical method (Whiting, Wilson, & Reeder, 2011), which required observation and mathematical skills, to examine a soil sample. The mechanical method required the use of a Mason jar, a 50-gram soil sample, water, powdered laundry detergent, and ruler. The soil, 500 mL of water, and detergent were added to the Mason jar and shaken, creating a soil suspension. The Mason jar was placed on the table, and the children made initial observations. On the second day of the soil lesson, the children observed how many layers settled out of the suspension and measured each layer using a ruler. A mathematical formula was provided for older children to calculate the percentages of sand, silt, and clay (see Appendix D).

Field Trips. In this chapter, we focus on the field trip to Rocky Mountain National Park because it aligned more closely with the natural environment. Park rangers led the activities during the field trip. Interns and students had a chance to examine animal pelts and use clues to discover which animal preyed on another using the Crime Scene Investigation (CSI) scenario. Activities also included a nature walk (i.e., neighborhood watch) that included observations of fungi, bacteria, and invertebrates (i.e., FBI) and pine trees infested with beetles. The final activity investigated how the valleys were formed by simulating the forces of glaciers (i.e., pushing ice), rain (i.e., sprinkling water), rivers (i.e., pouring water), and rocks (i.e., dropping small pebbles) in a bin of soil.

Interns' Learning and Preparation for Teaching

The interns in the summer camp had stronger content knowledge, in general, than interns in the Saturday academies due to the researchers' purposeful solicitation and careful screening of applications. Summer interns observed and facilitated learning in small groups of two or three students. Field notes revealed they demonstrated activities instead of working alongside students. However, excerpts from questionnaires also reveal the interns enjoyed learning natural science. They had positive comments about the activities at the park:

> *What stands out for me with my work this week was going to the Rocky Mountain Park. It stood out for me because the kids were genuinely excited about the lessons they received.*
> *When we took the trip to Rocky Mountain National Park...one of the children I got to observe and work with, enjoyed not only just being outside but how he got to be hands-on with many things we did.*
> *Students also seemed to really enjoy learning at the mountains; looking at the animal pelts.*

The foregoing evidence suggested both the interns and the student participants enjoyed being outdoors and learning natural science. One intern was enthralled with animal pelts. He smiled as he explored the pelt and its uses by creating a Superman-like cape.

Student participants also confirmed their enjoyment at Rocky Mountain National Park during focus group interviews:

> *Yesterday, when we went to [the] national park.... The neighborhood watch, [I liked] how we did the hike and looked at different trees.*
> *We went to Rocky Mountain National Park. That was fun...I like hiking, and I like the mountains more than I like the city. So that was fun. And I learned a couple of things about certain animals that I didn't know.... I learned about...wolverine. I [didn't]...know wolverine....*

However, two student participants did not appreciate some of the interns' teaching methods:

> *... I really didn't like how [some of the interns] wanted to always be in charge of everybody. And I'm not saying my intern was like that, but most of the interns, when I*

> was watching, they would act like they would be in charge of everybody, and they were actually not. And there was one intern that just wanted to be in charge of a table, and I kind of felt bad for those students, because they really didn't get to do anything but their work.
> Some stuff was kind of boring like waiting and just watching the interns do the stuff instead of the kids doing it. Yeah, so...we could be more interactive.

One of the interns concurred that less supervision was more desirable in this teaching-learning context. A second intern believed he was underutilized while another intern commented on the autonomy and responsibility she felt during small group instruction:

> I think the kids would also enjoy doing some activities with less intern supervision by letting them handle objects on their own.
> I feel like I was just used as a glorified babysitter. Myself, I felt like I wasn't utilized to my fullest potential. There should be more focus on the actual learning than activities and experiments.
> I really like that we were given a lot of freedom when working with the children and that we were trusted with such a great responsibility.

Interns' Interactions With Students

We analyzed data obtained from the summer interns' questionnaires and focus group interviews with interns and student participants to understand how interns interacted with and engaged children in STEM activities. The microscope activity provided the interns with an opportunity to demonstrate and teach biology concepts. Field notes revealed interns and student participants had a great deal of interaction during the microscope activity. The interns guided the students as they made slides and demonstrated how to use sunlight and the mirror to see the slide under the microscope. An excerpt from one of the interns' questionnaire revealed children enjoyed this novel activity:

> One of the activities I saw the kids really enjoy was working with the microscopes. For some this was the first time they got to handle them.

Student participants confirmed their enjoyment and engagement during this activity:

> *What was cool for me was when we looked at the cells....*
> *A new thing for me was also the cells. Mainly at school we were looking at different stuff.*

Another example of interaction and STEM learning took place during the soil analysis lesson. One of the interns was videotaped as he explained the procedure and the mathematics needed to complete the soil analysis. We use a pseudonym to describe his brief interaction with students in his small group. The verbatim transcript of his explanation is as follows:

> Oscar: You put something that you want and something that you need on the other side. So, for example we have sand. You put 22 cm of sand and you have 5.01 cm... (so) for the total of the whole jar you times it by 100. So what you can do, you put 22 times 100 divided by 5.01 and that will be your percentage of how much sand makes up the dirt. Does that make sense to everybody?
> Students: Y-e-s.

Oscar helped the students to understand how to calculate the percentage of sand in the soil mixture. Students appeared to understand his instructions and utilized procedural knowledge to complete the soil analysis without much difficulty. This lesson revealed the importance of knowing mathematics in order to use it as a tool in science.

Emergent Themes. Using the constant-comparison method and open coding (Strauss & Corbin, 1990), two themes emerged during the microscope and soil analysis lessons: *connections, enjoyment,* and *student autonomy* or *independence*. Oscar realized that some students needed stronger connections to the content to understand why the activities and experiments worked. While science concepts were explained at the outset and the interns presented the demonstrations, perhaps all student participants did not clearly understand the concepts. Making explicit connections between the activities and the content may have deepened students' understanding of the lessons. Successfully engaging students in content is an important tenet of CRP (Howard, 2003; Ladson-Billings, 2009).

A second theme that emerged from the data was enjoyment. Students expressed enjoyment and excitement about the lessons and activities provided at the national park. These activities included learning about predators like wolverines, examining animal pelts, and taking nature walks. Many of the urban students who participated in the summer camp had limited experience with wildlife. Learning about wolverines was a surprise. One student thought they were only fictitious characters. The interns also enjoyed the national park. One was so intrigued with animal pelts that he modeled it like a cloak or cape.

The final theme was student autonomy or independence. Children wanted less supervision, and one of the interns agreed that they should have greater autonomy during the activities. This was evident primarily during the microscope lesson. One factor that may have influenced less autonomy during this lesson was the cost of the microscopes and the slides. Many of the students had not used microscopes before. It was understandable for staff and interns to monitor their use, especially with slides and slide covers, which were made of glass.

Using open coding, the theme that emerged during the field trips was *exposure to outdoor education*. Exposing underrepresented children and youth to place-based field experiences, reinforced appreciation for natural science. Given the 100th anniversary of the National Park Service on August 25, 2016 <http://www.nationalparks.org> and outreach efforts like Latino Outdoors <http://latinooutdoors.org/>, exposing underrepresented students to the natural environment increased awareness of conservation and protecting our natural resources. Interns explored and learned alongside the children by engaging in place-based activities at the national park. The foregoing excerpts reveal the influence of informal learning environments and experiential STEM learning as the power dynamics amongst the interns and the students were greatly reduced and, in many instances, they became peers who were interested in learning about various topics (Puvirajah, Verma, & Webb, 2012).

Interest in STEM Fields or Teaching

Given the short duration of the summer camp, we did not administer the STEM Semantics survey to interns in this context. However, the interns' questionnaires were reviewed to ascertain their interest in STEM/STEM-related fields and teaching careers. Eleven of the interns who worked during the summer internship completed the questionnaire and responded to the following item: "When I am older, the kind of work I would like to be doing is...." However, three of the summer interns had also participated in the Saturday academies. Therefore, their questionnaires were removed from the sample. Analyses of the new respondents reveal that six (75%) of the remaining eight interns were interested in STEM or STEM-related careers (i.e., two health science; one engineering; one aviation; one architecture; and one pathology) and one was interested in teaching (13%). Given that four (50%) of these interns were STEM or education majors at the outset; increased STEM interest was a positive outcome. However, there appeared to be no change in teaching interest.

> I am an education major so this teaching internship interested me. I applied because I felt I could make a strong impact on the summer internship.

Interestingly, data obtained from focus group interviews with four target interns revealed two additional interns had interests in teaching that were not captured by the questionnaire.

> Since eighth grade, I've always wanted to be a teacher...I got to practice my skills. Well I want to be an elementary teacher and math and science are two of my most favorite subjects so this kind of gave me the opportunity to learn even more about it and so it's something I would have to continue.

By matching the ID numbers, we learned these two interns had selected another career on the questionnaire. Nevertheless, the data reveal they were open to teaching. If we include these two interns' interest in teaching, 38% were interested in teaching after working in the summer internship.

Discussion. The results of this study reveal three findings: (a) the quality of interns' training and field-based experiences differed by the type of context; (b) the interns' engagement with students was dependent on their content knowledge and exposure to nature and science; and (c) 75% of the interns in this study developed or maintained prior interest in STEM/STEM-related careers or teaching.

In regard to the training, while both cohorts had four hours of training prior to the internship, the number of hours they worked with students varied a great deal. Interns in the Saturday academies worked two-and-a-half hours per session over four weeks while summer interns worked six hours per day for one week. Thus, the summer interns needed much more training to prepare them to teach than the Saturday interns. Furthermore, the data reveal the importance of communicating clear expectations prior to the field experience so that interns understand their roles and responsibilities. The interns in the Saturday academies were well prepared by the geologist with concrete lesson plans on paleontology. As a result, they had successful field experiences. While the summer interns were provided with the lesson plans, there were six topics covered during the weeklong summer camp. In comparison to 10 contact hours on dinosaurs, these interns taught three times as many topics during 30 contact hours. Thus, as Oscar observed, the lessons in the summer camp appeared to be a hodge-podge of activities and experiments rather than deep learning.

The summer teaching experiences could have been further illuminated if the researchers had focused on fewer topics and allowed the interns to break down challenging science concepts during the summer camp. Similar to Vygotsky's (1978) Zone of Proximal Development (ZPD), the soil scientist and the researchers should have provided more training on selected topics, allowed the interns to observe and shadow them at the beginning of the week, and then gradually take on more teaching responsibility toward the middle and the end of the week. In this way, cultural relevance and place-based learning could have been made more apparent as well. For the most part, culture and place were missing from the lessons presented at the community center during the summer camp, as children did not articulate what they learned and how it applied to their lives. While the purpose of the summer camp was to offer students rich science lessons embedded in culture and place, it is also a cautionary tale about good intentions gone awry.

In terms of STEM engagement, results show that the interns' interactions with participating students were dependent upon their content knowledge and exposure to nature and science. Some of the interns in the Saturday academies had weak science content knowledge. Thus, the themes of discovery and collaborative learning emerged in this context. While Saturday interns had concrete lesson plans, they admitted to learning alongside the students. Yet, they were deeply involved in the teaching-learning process because they were learners themselves. Four summer interns (33%) had sufficient STEM content knowledge, as evident by Oscar's teaching episode. Yet, when it came to the field trips, most of these interns also became learners as several of them experienced nature and wildlife for the first time.

Prospective education majors have few opportunities to engage in outdoor field experiences to learn science (Capraro *et al.*, 2010). Low rates of participation are especially prevalent in geology and environmental science (National Science Foundation, 2010). Yet, the internship contributed to new and sustained interest in STEM fields and/or teaching (Borgerding, 2015). However, none of the interns mentioned an interest in STEM education explicitly, even though some were biology and mathematics majors. Moreover, males are underrepresented in the teaching field. The majority of the interns in this study were male. However, only two males expressed an interest in teaching. Further research on high school and college-aged students is needed to determine the impact of internships and other factors on recruiting STEM majors into education (Borgerding, 2015; Leonard *et al.*, 2011). Teachers with strong science and mathematics content knowledge are needed in K–12 schools (Bishoff *et al.*, 2014).

Conclusions

While significant effort has been made to broaden and extend opportunities for underrepresented students to engage in and be exposed to STEM learning, persons of color remain underrepresented in STEM careers (National Academy of Sciences *et al.*, 2011). This may not be a result of a lack of preparedness inasmuch as it is a lack of opportunities to develop as STEM experts. It is important to note that the interns who interacted with underrepresented students in this study were also from communities of color. As young adults in these communities, they are often judged by their racial and ethnic identities rather than their science identity (Carlone & Johnson, 2007; Wallace & Brooks, 2015). Their voices and experiences are important, particularly when they serve as interns and role models for children of color (Bracey, 2013). Moreover, the opportunities to serve as a mentor enabled young people of color to develop their STEM identity, especially as STEM educators. In fact, several of the interns have since pursued vocations in STEM education.

While there is research that suggests children learn more from instructors who share their background (Hrabowski & Sanders, 2015; Waddell & Ukpokodu, 2012), others report that race and/or ethnicity do not necessarily correlate with teacher quality (Sleeter & Thao, 2007). However, teacher quality is a reflection of teacher training and preparation. Despite the limitations, the interns in this study were able to model, explain, and teach science content (in varying degrees) to children during small group instruction.

With such positive reports from the investigation, one may be left wondering, why the high degree of success. It is hypothesized, though not necessarily substantiated with empirical data, that several talents and characteristics that interns brought to the experience helped to establish them as science apprentices. First, background characteristics of the interns were varied, and they served as role models to young learners. Second, by anyone's definition, the interns demonstrated impeccable personal character. Furthermore, nearly all of the interns and many of the individuals delivering the coursework were people of color, and it most likely helped urban students, who had the opportunity to see positive role models who grew up in a similar environment, to excel in the program. It is one thing to tell young learners that they can accomplish anything. Such comments can be patently patronizing and tenuous. It is another thing altogether to utilize peers of the same background, with a high degree of expertise, to work with the young learners. As an example, one of the senior personnel identified as an African American. She had earned a PhD in microbiology, worked for the U.S. Department of Agriculture, and was actively involved with students on a regular basis. In this respect, she was a positive STEM role model for the young students who participated in the study as well as the interns.

The high school and college interns were also individuals who were able to support children's learning and simultaneously hold high expectations of them. The cornerstone of the entire experience was clear communication of very high expectations. Senior personnel delivering the courses communicated high expectations to mentors, and mentors communicated high expectations for young learners. Hence, the high degree of success did not serendipitously occur. Instead, senior personnel and mentors maintained a high level of quality control that ultimately resulted in a highly successful experience as planned.

Recruiting interns to teach culturally relevant, place-based science is not a panacea. Providing children, youth, and young adults of color with a *third space* (Lipka et al., 2009), where they have the freedom to discover and voice their understanding leads to cultural competence and empowerment. Critical race theory supports hearing the voices of marginalized students in science and other fields that provide access to quality education and careers (Delgado & Stefancic, 2012). It is therefore critical to note that interns in this study positively influenced the student participants and most importantly the researchers who learned that interns are more than glorified babysitters.

Although the results do not conclusively illustrate deep learning on behalf of the interns and student participants, the results are encouraging. The interns' collaborative approach toward science and teaching facilitated learning among student participants that is not generally experienced in traditional classrooms (Leonard et al., 2011). These millennials were novice instructors and mentors who taught and nurtured their younger counterparts. Coupled with place-based learning in novel contexts, these experiences affirmed the interns' interest in STEM fields and teaching. The results of this study support the notion that a STEM career begins with opportunities to learn STEM (Sadler et al., 2012). Likewise, the path to STEM teaching is making deep connections to content in culturally relevant, place-based learning environments.

Acknowledgments

The material presented in this chapter is based upon work supported by the National Science Foundation, award number GEO #1260957. Any opinions, findings, and conclusions or recommendations expressed in this publication are those of the author(s) and do not necessarily reflect the views of the National Science Foundation.

Note

1. Gender neutral term for Latina/o.

References

Aikenhead, G. S. (1996). Science education: Border crossing into the subculture of science. *Studies in Science Education, 27*, 1–52.

Aikenhead, G. S., Calabrese Barton, A., & Chinn, P. W. U. (2006). Forum: Toward a politics of place-based science education. *Cultural Studies of Science Education, 1*, 403–416.

Akom, A. (2011). Eco-apartheid: Linking environmental health to educational outcomes. *Teachers College Record, 113*(4), 831–859.

Anderson, L. W., & Bourke, S. F. (2000). *Assessing affective characteristics in schools* (2nd ed.). Mahwah, NJ: Lawrence Erlbaum Associates.

Bell, D. A. (1980). Brown v. Board of Education and the interest convergence dilemma. *Harvard Law Review, 93*(3), 518–533. doi: 10.2307/1340546.

Bishoff, P., French, P., & Schaumloffel, J. (2014). Reflective pathways: Analysis of an urban science teaching field experience on Noyce Scholar-Science Education awardee's decisions to teach science in a high-need New York City School. *School Science and Mathematics, 114*(1), 40–49.

Borgerding, L. A. (2015). Recruitment of early STEM majors into possible secondary science teaching careers: The role of science education summer internships. *International Journal of Environmental & Science Education, 10*(2), 247–270.

Bourdieu, P. (1973). Cultural reproduction and social reproduction. In R. Brown (Ed.), *Knowledge, education and cultural changes* (pp. 56–69). London: Tavistock.

Bracey, J. (2013). The culture of learning environments: Black student engagement and cognition in math. In J. Leonard & D. B. Martin (Eds.), *The brilliance of Black children in mathematics: Beyond the numbers and toward new discourse*, (pp. 171–194). Charlotte, NC: IAP.

Capraro, M. M., Capraro, R. M., & Helfeldt, J. (2010). Do differing types of field experiences make a difference in teacher candidates' perceived level of competence? *Teacher Education Quarterly, 31*(1), 131–154.

Carlone, H. B., & Johnson, A. (2007). Understanding the science experiences of successful women of color: Science identity as an analytic lens. *Journal of Research in Science Teaching, 44*(8), 1187–1218.

Cochran-Smith, M. (2004). The multiple meanings of multicultural teacher education. A conceptual framework. In F. Schultz (Ed.), *Annual editions: Multicultural education* (pp. 26–35). Guildford, CT: McGraw-Hill/Dushkin.

Creswell, J. W. (1998). *Qualitative inquiry and research design choosing among five traditions*. Thousand Oaks, CA: Sage Publications.

DeJarnette, N. K. (2012). America's children: Providing early exposure to STEM (Science, Technology, Engineering and Math) Initiatives. *Education, 133*(1), 77–84.

Delgado, R., & Stefancic, J. (2012). *Critical race theory: An introduction*. New York City, NY: New York University Press.

Dewey, J. (1916). *Democracy and education: An introduction to the philosophy of education*. New York, NY: Macmillan.

Evans, B. R., & Leonard, J. (2013). Recruiting and retaining Black teachers to work in urban schools. *Sage Open*, 1–12. doi: 10.1177/2158244013502989.

Gay, G. (2010). *Culturally responsive teaching: Theory, research, and practice*. New York, NY: Teachers College Press.

Greer, B., Mukhopadhyay, S., Powell, A. B., & Nelson-Barber, S. (Eds.). (2009). *Culturally responsive mathematics education*. New York, NY: Routledge.

Gruenewald, D. A. (2003). Foundations of place: A multidisciplinary framework for place-conscious education. *American Educational Research Journal, 40*(3), 619–654. doi: 10.3102/00028312040003619

hooks, B. (2009). *Belonging: A culture of place*. New York, NY: Routledge.

Howard, T. C. (2003). Culturally relevant pedagogy: Ingredients for critical teacher reflection. *Theory into Practice, 42*(3), 195–202.

Howley, A., Showalter, D., Howley, M. D., Howley, C. B., Klein, R., & Johnson, J. (2011). Challenges for place-based mathematics pedagogy in rural schools and communities in the United States. *Children, Youth and Environments, 21*(1), 101–127.

Hrabowski, F. A. III, & Sanders, M. G. (2015). Increasing racial diversity in the teacher workforce: One university's approach. *Thought & Action, 31*(2), 101–116.

Knezek, G., & Christensen, R. (2008). STEM Semantics Survey (v. 1.0).

Ladson-Billings, G. (1995). Toward a theory of culturally relevant pedagogy. *American Educational Research Journal, 32*(3), 465–491.

Ladson-Billings, G. (2005). *Beyond the big house: African American educators on teacher education*. New York, NY: Teachers College Press.

Ladson-Billings, G. (2009). *The dreamkeepers: Successful teachers of African American children* (2nd ed.). San Francisco, CA: John Wiley & Sons.

Leonard, J., Barnes-Johnson, J., Dantley, S. J., & Kimber, C. T. (2011). Teaching science inquiry in urban contexts: The role of elementary preservice teachers' beliefs. *The Urban Review, 43*(1), 124–150.

Leonard, J., Chamberlin, S. A., Johnson, J. B., & Verma, G. (2016). Broadening urban students' opportunities to learn science and influencing student interest through place-based education. *The Urban Review, 48*(3), 355–379. doi: 10.1007/s11256-016-0358-9.

Leonard, J., & Oakley, J. E. (2006). We have lift off! Integrating space science and mathematics in elementary classrooms. *Journal of Geoscience Education, 54*(4), 452–457.

Lewis, R. B., & Maas, S. M. (2007). QDA Miner 2.0: Mixed-model qualitative data analysis software. *Field Methods, 19*(1), 87–108.

Lipka, J., Yanez, E., Andrew-Ihrke, D., & Adam, S. (2009). A two-way process for developing effective culturally based math. In B. Greer, S. Mukhopadhyay, A. B. Powell, & S. Nelson-Barber (Eds.), *Culturally responsive mathematics education* (pp. 257–280). New York, NY: Routledge.

Lucas, T., & Villegas, A. M. (2011). A framework for preparing linguistically responsive teachers. In T. Lucas (Ed.), *Teacher preparation for linguistically diverse classrooms: A resource for teacher educators* (pp. 55–72). New York, NY: Routledge.

Maranto, R., & Shuls, J.V. (2012). How do we get them on the farm? Efforts to improve rural teacher recruitment and retention in Arkansas. *The Rural Educator, 34*(1), Retrieved from http://epubs.library.msstate.edu/index.php/ruraleducator/article/view/138.

Martin-Hansen, L., Puvirajah, A., & Verma, G. (2012). Creating a pipeline to STEM careers through service learning: The AFT program. In R. E. Yager (Ed.), *Exemplary science careers in science and technology* (pp. 111–128). Arlington, VA: NSTA Press.

Monk, M. H., Baustian, M. M., Saari, C. R., Welsh, S., D'Elia, C. F., Powers, J. E., … Francis, P. (2014). EnvironMentors: Mentoring at-risk high school students through university partnerships. *International Journal of Environmental & Science Education, 9*(4), 385–397.

Moore, C. M. (2012). The role of school environment in teacher dissatisfaction among U.S. public school teachers. *SAGE Open, 2*, 1–16. doi:10.1177/215824401243888

National Academy of Sciences, National Academy of Engineering, and the Institute of Medicine. (2011). *Expanding underrepresented minority participation: America's science and technology talent at the crossroads*. Washington, DC: The National Academies Press.

National Science Foundation. (2010). *Preparing the next generation of STEM innovators: Identifying and developing our nation's human capital*. Retrieved from http://www.nsf.gov/nsb/publications/2010/nsb1033.pdf

Puvirajah, A., Verma, G., & Webb, H. (2012). Examining the mediation of power in a collaborative community: Engaging in informal science as authentic practice. *Cultural Studies of Science Education, 7*(2), 375–408.

Rye, J. A., Selmar, S. J., Pennington, S., Vanhorn, L., Fox, S., & Kane, S. (2012). Elementary school garden programs enhance science education for all learners. *Teaching Exceptional Children, 44*(6), 58–65.

Sadler, P. M., Sonnert, G., Hazari, Z., & Tai, R. (2012). Stability and volatility of STEM career interest in high school: A gender study. *Science Education, 96*(3), 411–427.

Sleeter, C., & Thao, Y. (2007). Guest editors' introduction: Diversifying the teaching force. *Teacher Education Quarterly, 34*(4), 3–8.

Solorzano, D. G., & Yosso. T. J. (2002). Critical race methodology: Counter-Story telling as an analytical framework for education. *Qualitative Inquiry, 8*, 23–44.

Strauss, A. L., & Corbin, J. (1990). *Basics of qualitative research: Grounded theory procedures and techniques*. London: Sage.

United States Department of Agriculture, Natural Resource Conservation Service (USDA-NRCS). (2005). *Urban soil primer: For homeowners and renters, local planning boards, property managers, students and educators*. Washington, DC: U.S. Govt. Printing Office. Retrieved from http://www.ncrs.usda.gov/Internet/FSE_Documents/nrcs142p2_o52835.pdf

United States Department of Education Office of Post-Secondary Education. (2013). *Preparing and credentialing the nation's teachers: The secretary's ninth report on teacher quality*. Washington, DC: U.S. Department of Education, Office of Postsecondary Education. Retrieved from https://title2.ed.gov/TitleIIReport13.pdf

Verma, G., Puvirajah, A., & Webb, H. (2015). Enacting acts of authentication in robotics competition: An Interpretivist study. *Journal of Research in Science Teaching, 52*, 268–295. doi: 10.1002/tea.21195

Vilson, J. L. (2015). The needs for more teachers of color. *American Educator, 39*(2), 27–31.

Vygotsky, L. (1978). *Mind in society: The development of higher psychological processes*. Cambridge, MA: Harvard University Press.

Waddell, J., & Ukpokodu, O. N. (2012). Recruiting & preparing diverse urban teachers: One urban-focused teacher education program breaks new ground. *Multicultural Education, 20*(1), 15–22.

Wallace, C. S., & Brooks, L. (2015). Learning to teach elementary science in an experiential, informal context: Culture, learning, and identity. *Science Education, 99*(1), 174–198.

Whiting, D., Wilson, C., & Reeder, J. (2011). Estimating soil texture: Sand, silt or clayey? Colorado Master Gardener Program Garden Notes #214. Colorado State University Extension. Retrieved January 2016, from http://www.ext.colostate.edu/mg/gardennotes/214.html

Worsham, H. M., Friedrichsen, P., Soucie, M., Barnett, E., & Akiba, M. (2014). Recruiting science majors into secondary science teaching: Paid internships in informal science settings. *Journal of Teacher Education, 25*(1), 53–77.

Developing Pre-Service Mathematics Teachers to Meet the Needs of Black Male Students in Teacher Education Programs

JULIUS DAVIS, RAMON B. GOINGS, AND KEISHA M. ALLEN

Abstract

Preparing preservice mathematics teachers to teach Black male learners who are characterize as a problem in society and mathematics spaces present challenges for faculty in teacher education programs. These faculty members must address, examine, and reframe pre-service mathematics teachers' perspectives of Black male students to ensure their needs are being meet. In this chapter, we call for mathematics teacher educators to take their role seriously in helping pre-service mathematics teachers see the assets and strengths of Black male students as humans, students, and mathematics learners. We contend that mathematics teacher education programs must prepare pre-service teachers to nurture Black male students' mathematical brilliance in the classroom. This chapter offers recommendations on the changes needed in mathematics teacher education programs to prepare pre-service teachers to meet the needs of their Black male students.

Introduction

Through biased and racist imagery, Black male learners have been negatively portrayed in the classroom and beyond (Howard, 2013; Noguera, 2009; Polite & Davis, 1999). Black boys and young men are often depicted as barely human, criminals, unintelligent, and inferior, all of which support standing racist assumptions that have been used to justify inhumane treatment toward them (Howard, 2013; Noguera, 2009; Polite & Davis, 1999). Unfortunately, these societal characterizations shape the views of those responsible for their schooling and the style of interacting employed by teachers, principals, and other school administrators (Howard, Flennaugh, & Terry, 2012). Black male students receive high rates of disciplinary infractions, are suspended and expelled at high rates, are consistently reported as low achievers in all subject areas, are overrepresented in special education classrooms, and are underrepresented in gifted and talented programs (Goings & Ford, 2018; Howard, 2013). These educational reports have reinforced racist societal views that operate at both a conscious and subconscious level. Howard et al. (2012) warned about the impact of social imagery because it "becomes an integral part of a population's thinking when it is institutionalized for a sustained period of time through different venues, and shapes generations of people's thinking about a particular reality or perceived reality" (p. 85). Sadly, the field of education is an institutional venue where negative social imagery about Black male students is maintained and reified.

Mathematics teacher educators and pre- and in-service teachers are not exempt from the effects of the negative societal images and perspectives about Black male students. Reports of Black male students' overrepresentation in lower level and remedial mathematics courses; underrepresentation in higher level, college preparatory courses; and low mathematics performance via standardized test scores and course grades, have reinforced societal views about their educability (Berry, 2008; Davis, 2014; Oakes, 2005; Polite, 1999; Thompson & Lewis, 2005). These reports have contributed to Black male students being characterized as incapable, lacking natural mathematics ability and the capacity or interest to learn (Martin, 2007). These characterizations shape the views, beliefs, and perspectives of teacher educators and pre- and in-service teachers. Inevitably, their teaching, thinking, and engagement are affected.

Given this reality, we argue that mathematics teacher education programs and professional development must address, examine, and reframe teachers' perspectives to better attend to the needs of Black male students. We call for mathematics teacher educators to take their role seriously in helping pre- and in-service teachers see how societal imagery and perspectives subconsciously shape their views and expectations of Black males as humans, students, and mathematics learners. We contend that teacher education programs must prepare teachers to nurture Black male minds and see the strengths they bring to the classroom, notably their mathematical brilliance.

To achieve the aim of this chapter, we first contextualize this issue by discussing two metanarratives that have been used as lenses to examine and study Black male students in mathematics education. We then discuss the research on mathematics teacher education programs, and the lack of focus on issues related to race, racism, gender, class, power, and privilege in mathematics education. Lastly, we conclude with recommendations on the changes needed in mathematics teacher education programs, so teachers are adequately prepared to meet the needs of their Black male students.

Examining Deficit- and Strength-Based Perspectives of Black Male Students in Mathematics Education

The two main lenses used to discuss Black male students' mathematics education are those of *achievement* and *experience* (Martin, 2007). The achievement lens is intricately connected to the dominant narrative, while the counternarrative is more closely aligned to the experience lens. The dominant narrative provides a deficit perspective of Black boys and young men that suggests an abnormality, an aberrance, among them and their families. It suggests that they lack the ability to do and be successful in mathematics (Berry, 2008; Davis, 2014; Martin, 2007). The achievement lens privileges standardized test outcomes, leading to narrow definitions of teaching and learning. It explains how meritocracy and a hard work ethic are values Black male students should aspire to achieve if they want to perform at the same levels as their White male peers (Davis, 2014; Martin, 2007; Stinson, 2008, 2011, 2013). The achievement lens also promotes the idea and belief "that some people, or groups, have natural math ability" (Martin, 2007, p. 14). According to this lens, Black male students do not have the ability as reflected in their standardized mathematics performance (Davis, 2014).

The counternarrative (i.e., experience lens) focuses on the strengths of Black male learners as well as high-achieving and successful Black male students in mathematics education. It also challenges the dominant narrative which foregrounds Black male students' failure and low achievement. The experience lens seeks to broaden mathematics teacher educators' and teachers' perspectives of the knowledge, skills, and dispositions (Martin, 2007) needed to teach Black male students. It provides deeper insights into Black male students' achievement by centering the ways they experience mathematics teaching, learning, participation, and access through their racialized and gendered identities (Jett, 2011; Martin, 2007).

The Achievement Lens

The achievement lens positions achievement, narrowly defined as test scores, as the major criterion of worth in education. Student achievement becomes the criterion of teacher effectiveness, the determinant of children's ability and intelligence and the indicator of successful school leadership. Through this lens a successful student is one with high test scores, a minimum 3.0 GPA, having proficient or above standardized test scores, and taking and succeeding in college preparatory or advanced mathematics courses. It is this lens that gives rise to the notion of the achievement gap. Martin (2009a) argues that the so-called racial achievement gap supports the notion that some groups, mainly White and Asian males, possess mathematical ability and represent the standard of competency. Black male students, on the other hand, are perceived as problematic and unable to perform at high levels. Martin (2007) also argues that the discourse surrounding the so-called racial achievement gap positions Black male students as deficient, mathematically illiterate, and less-than-ideal learners.

The pervasive discourse surrounding Black male students' lack of mathematical ability derives from this view of education and gives rise to the often misguided efforts to increase test scores. This effort has led to the phenomenon of *teaching to the test*, where Black male students receive rigid, nondynamic instruction that focuses only on test scores rather than conceptual and procedural knowledge (Davis, 2014; Lattimore, 2003, 2005a, 2005b). Davis and Martin (2008) define teaching to the test as "classroom practices that emphasize remediation and skills-based instruction over critical and conceptual-oriented thinking, decreased use of rich curriculum materials, narrow teacher flexibility in instructional design and decision making, and the threat of sanctions for not meeting externally generated performance standards" (p. 11). Some mathematics teachers use teaching to the test as the dominant instructional approach for Black male students (Davis, 2014; Lattimore, 2005a, 2005b). For Black students this instructional approach coupled with low teacher expectations inhibits the construction of deep conceptual understanding and the development of procedural fluency in mathematics, which is needed for higher level, college preparatory, honors, gifted and talented, and advanced mathematics courses and programs (Davis, 2014; Lattimore, 2005a, 2005b; Polite, 1999).

Although Martin (2007) does not frame mathematics course-taking patterns through the achievement lens, we contend that Black male students' mathematics course-taking patterns are the result of the achievement perspective, which focuses on their perceived lack of mathematical ability. Black male students are overrepresented in lower level and remedial mathematics courses and underrepresented in higher level, college preparatory, honors, gifted and talented, and advanced mathematics courses and programs (Berry, 2008; Noble, 2011; Oakes, 2005; Polite, 1999; Thompson & Davis, 2013; Thompson & Lewis, 2005). Moreover, the pervasive discourse surrounding Black male students' lack of mathematical ability becomes a self-fulfilling prophecy and results in many of these previously mentioned inequities (e.g., overrepresentation in the lower level and remedial mathematics courses).

Many factors explain why Black male students lack access to specialized mathematics courses and programs. Stakeholders using the achievement lens tend to ignore these factors. The experience lens reveals many reasons for Black male students' lack of access and participation in specialized mathematics courses and programs. Low teacher expectations, poor quality instruction, teaching to the test, standardized test results, and achievement gap discourse in mathematics are some of the many reasons used to explain Black male students' lack of access to and participation in specialized mathematics courses and programs (Berry, 2008; Davis, 2014; Thompson & Lewis, 2005).

The Experience Lens

By contrast, the experience lens, places the classroom and school experiences of learners as the focal point in attempts to understand educational outcomes. This lens takes as its underlying premise the idea that all children can learn and be successful, and that it is the pedagogical approaches to which they are exposed and the social and cultural environment of the school, the home and society that are the primary determinants of school outcomes. This lens also brings into view the need to provide experiences that are suited to the varying needs of children from differing racial, gender, ethnic, cultural, and ability groups. From this perspective, some researchers studying the education of Black male students have begun to focus on the school experiences of this group of students. The majority of research focused on Black male students' experiences has centered on students who are deemed to be high-achieving and successful learners (Berry, 2008; Hrabowski, Maton, & Grief, 1998; Jett, 2009, 2011, 2016; McGee, 2013, 2014; McGee & Martin, 2011; Terry & McGee, 2012; Thompson & Davis, 2013; Thompson & Lewis, 2005). However, some mathematics education researchers have not focused solely on high-achieving and successful Black male students. Studies by Corey and Bower (2005), Davis (2014), Lattimore (2005b), and Terry (2010, 2011) have included Black male students who are not considered high achievers or who have not always performed at high levels in mathematics. The experience lens has provided deeper insights into varying Black male students' experiences with mathematics teaching, learning, and participation through their racialized and gendered lens.

Research on high-achieving and successful Black male students has highlighted support (from family, caregivers, extended family, community, mathematics teachers, and peers) as a key factor in mathematics success (Berry, 2008; McGee, & Pearman, 2014; Stinson, Jett, & Williams, 2013;

Terry & McGee, 2012; Thompson & Davis, 2013). Not all high-achieving and successful Black male students have all of these supports, but they all have some type of support. It is also important to note that some Black males have these supports and still do not perform at high levels. However, overall, academic and emotional supports are helpful for those who do.

High-achieving Black male students tended to have mathematics teachers who have high expectations and are available, flexible, supportive, and encouraging. A limitation of research reporting mathematics teacher effectiveness for Black male students is the absence of findings that provide insight into the teachers' mathematical content knowledge (Martin, 2007). In Thompson and Lewis' (2005) case study of a Black male student, the mathematics teacher's knowledge of calculus, appeared to be insufficient.

Research on Black male students in mathematics education has revealed that only some have access to higher level, college preparatory, honors, gifted and talented, and advanced mathematics courses and programs (Berry, 2008; Polite, 1999; Terry & McGee, 2012; Thompson & Lewis, 2005). McGlamery and Mitchell (2000) provide insight into teacher-led efforts to increase Black male students' participation in upper-level mathematics courses through cohort recruitment models, modified classroom environments, role-model and career exposure, and increased support in and outside of the classroom. However, the vast majority of Black male students do not gain access to these specialized courses and programs. For example, some Black male students did not gain access to such programs because their schools lacked the resources to offer them (Ladson-Billings & Tate, 1995; Thompson & Lewis, 2005). Other reasons for the exclusion of Black males from the upper-level courses included poor advising (Polite, 1999). And sometimes teachers deliberately chose not to recommend Black males for nonacademic reasons such as personal struggles with the student (Berry, 2005). Some students were admitted into advanced mathematics course, but the quality of the mathematics course was substandard (Thompson & Lewis, 2005). Further, some qualified Black males elect not to enroll in higher level mathematics programs for varying reasons, including but not limited to personal or family financial constraints and responsibilities (Thompson & Davis, 2013).

The experience lens has also provided insight into how standardized mathematics assessments shape the type of mathematics instruction, participation, experiences, and access Black male students have in classrooms. Researchers have provided insight into how standardized mathematics test scores were used to provide or deny Black male students access to higher level mathematics courses (Berry, 2005; McGlamery & Mitchell, 2000; Polite, 1999). Some Black male students met the requirements (Berry, 2005; Noble, 2011) but were still denied access to these specialized classes (Berry, 2005; Polite, 1999). Not only are test scores used to deny Black males' access, but they also shape the type of mathematics instruction they receive. Some Black male students receive instruction that only teaches to the test without helping them develop a procedural and conceptual understanding of mathematics (Davis, 2014; Lattimore, 2003, 2005a, 2005b).

Terry (2010, 2011) recommends using mathematical counterstories as a pedagogical strategy to engage Black males in urban mathematics classrooms. Mathematics counterstories are anchored in critical race theory, social justice, and culturally relevant approaches to teaching and learning mathematics (Larnell, Bullock, & Jett, 2016; Pitts Bannister, Davis, Mutegi, Thompson, & Lewis, 2017; Terry, 2010). Terry (2011) contends that mathematical counterstories allow educators to address missed opportunities to reorient Black males to the usefulness of mathematics. Terry (2011) states, "Attention must be paid to the 'voice' of male African Americans and their narratives about (mathematics) learning" (p. 28). Further, Terry (2011) encourages mathematics educators to "look for mathematics" in interesting and engaging spaces and to mathematize situations that are significant to Black males. We contend that mathematics educators can actively engage Black male students in rigorous, conceptually rich mathematics by using mathematical counterstories (Terry, 2010). However, they must first reorient themselves to the liberatory nature of mathematics (Martin, 2009; Martin & McGee, 2009). Mathematical counterstories provide a pedagogical approach to engage and prepare Black males to be change agents in their communities. They create an opportunity for Black male students to develop their mathematics and racial identities as doers of mathematics (Terry, 2010, 2011).

Black Male Mathematics and Racial Identity Development

The individual and collective experiences of Black male students shape how they develop their mathematics and racial identities. Identity is an aspect of Black male students that is not often considered through the perspective of an achievement lens but identity development has been found to be an essential element of Black male students' learning outcomes (Berry & McClain, 2009; Berry, Thunder, & McClain, 2011; Nasir & Hand, 2008; Thompson & Davis, 2013). A focus on students' experiences has however provided insight into how Black male students have developed their mathematics and racial identities. Mathematics identity refers to

> the dispositions and deeply held beliefs that students hold about their ability to perform and participate effectively in mathematical contexts and to use mathematics to change the conditions of their lives. A mathematics identity encompasses a person's understanding of himself or herself and how they are seen by others in the context of doing mathematics. Therefore, a mathematics identity is expressed in narrative form as negotiated self, is always under construction, and results from the negotiation of our own assertions and the external ascriptions of others. (Martin, 2007, p. 41)

Black male students' mathematics identities do not develop in isolation. They are a facet of a student's overall identity and develops along with racial identity, which focuses on their views of race; and their positions in society, the broader community, school, and mathematics classrooms. Racial identity is also influenced by the meanings they associate with others' views of them.

A notable finding in research on Black male students across different mathematics achievement levels is that their race, class, and gender impact their performance and experiences (Berry, 2005; Corey & Bower, 2005; Davis, 2014; Howard, 2013). Black male students experience individual, institutional, structural, and systematic racism in the greater society as well as in urban, suburban, and rural schools and classrooms (Berry, 2005; Corey & Bower, 2005; Davis, 2014; Howard, 2013). There is limited research on Black male students in rural mathematics classrooms (Berry *et al.*, 2011; Smith & Cage, 2000). Much mainstream mathematics education research "[relies] on inadequate and impoverished approaches to race, racism, and racialized inequality" (Martin, 2009, p. 297) to describe how these issues impact Black male students. Issues of race and racial inequality are embedded in the many forms of racism that exist in society and its institutions (e.g., schools, mathematics classrooms, etc.). The institutional, structural, and systematic nature of how Black male students experience racism can be seen in the "widely accepted, and largely uncontested, racial hierarchy of mathematical ability" (Martin, 2009, p. 315) embedded in the so-called achievement gap discourse that has socially constructed Black male students as mathematically illiterate and less-than-ideal learners.

Most mathematics teacher education programs are created without consideration of the scholarship about Black male students. Given the highly publicized nature of Black male students' achievement and social standing, most higher education faculty members and pre- and in-service K–12 teachers are familiar with the dominant narrative of Black male academic failure and assumed deficiencies. Teacher educators and pre- and in-service teachers are less familiar with the narrative surrounding Black male students' success and high achievement in mathematics education or their experiences in mathematics classrooms and programs. An essential component of Black male student scholarship is the usage of knowledge gained to improve the mathematical experiences, achievement, participation, learning, and instruction of all Black male mathematics learners regardless of their achievement level.

Preparing Pre-Service Mathematics Teachers to Teach Black Male Students: An Examination of Mathematics Teacher Education Programs

Reports using the achievement lens (e.g., standardized test scores, course grades, course-taking patterns) often indicate that Black males have limited success in mathematics (Davis, 2014; Stinson, 2006), but the experience lens provides a more in-depth and broader context within which to understand their varying levels of achievement. One issue that comes to the fore when investigating learning outcomes through the experience lens is the teacher variable. Such investigations have shown that mathematics teacher education programs often fail to prepare pre-service teachers to address the needs and issues impacting Black male students' experiences, performance, and persistence. Black males are the most penalized and least served population in U.S. schools and mathematics classrooms (Davis, 2014; Howard, 2013). Given their academic, mathematical, and social standing, it is apparent that the mathematical fates of Black male students depend on the ability and willingness of teacher educators to prepare pre-service teachers to best support Black male students. However, findings from Adler, Ball, Krainer, Lin, and Novotna's (2005) synthesis show a lack of focus on Black male student needs.

In a survey of research on mathematics teacher education programs, Adler *et al.* (2005) examined nearly 300 publications and elaborated on 160. They presented four central themes about mathematics teacher education programs: (a) Most studies are small-scale qualitative research studies (e.g., case studies); (b) most research is conducted by teacher educators studying teachers in their program; (c) most research on mathematics teacher education programs occurred in the US; and (d) some critical questions have been answered while other vitally important questions remain unexamined. In Adler *et al.*'s (2005) study, no mathematics teacher education programs focused on Black male students. In fact, they found fewer studies on

> teachers learning to directly address inequality and diversity in their teaching of mathematics. We know far too little about teachers' learning to directly address inequality and diversity within their teaching of mathematics and here we include culture, gender, language, socio-economic status and mathematical background. (p. 376)

Adler and her colleagues' synthesis found that mathematics teacher education programs did not consider issues of race, class, and gender. However, it must be noted that the synthesis did not include information about teacher education programs at Historically Black Colleges and Universities (HBCUs), an area worthy of future research.

Critical Black scholars have produced two volumes focused on teacher preparation, the Black community, Black students, and Black teachers (Sealey-Ruiz & Lewis, 2011; Sealey-Ruiz, Lewis, & Toldson, 2014). These volumes provide insight into Black male students' education and the preparation of Black male teachers. While discussions of mathematics teacher education, teacher educators, and pre-service teachers were represented in the special issue (Sealey-Ruiz & Lewis, 2011) and book (Sealey-Ruiz et al., 2014), there were no specific studies focused on Black male students and teachers in mathematics education. Sheppard (2011) provides valuable insight into the preparation of Black female elementary pre-service teachers, but there were no Black males included in the study. Similarly, Brown, Davis, and Kulm (2011) presented findings from their National Science Foundation funded research exploring how a mathematics teacher education program can prepare teachers for diverse middle school classrooms, but there were no Black male teachers in the study nor was there a focus on Black male students.

In McGee's (2014) study of 13 Black pre-service teachers, the experiences of four Black males were included. Her findings shed light on the importance of Black men (fathers, uncles, brothers, etc.) helping teach and develop mathematical competency early in the lives of the teachers in her study. She also shares the successes and challenges that the pre-service teachers experienced in learning mathematics at the collegiate level and dealing with bias and racism among faculty members. McGee found that many of her participants lacked mathematical content knowledge due to various factors, including their racialized experiences in mathematics classrooms. They needed additional professional development to improve their content knowledge.

Milner, Pabon, Woodson, and McGee (2013) collaborated to discuss their experiences preparing pre-service teachers to teach Black male students. Because of the specific challenges these students face in the American public-school system, the researchers argue for the necessity of the explicit and proper education of mathematics teachers working with this population. We support this assertion and argue for racially specific preparation for mathematics teacher educators. Using the commonality of their own experiences, they call for teacher educators to emphasize the heterogeneity of Black males and prepare teachers to create educational experiences that are more responsive and relevant to their needs.

Though McGee is a mathematics teacher educator, her discussion of Black male students does not focus on mathematics education. Instead, she focuses on disrupting deficit perspectives of Black male students to help pre-service teachers see the strengths of the students and their families. Our conceptualization picks up where McGee left off and focuses specifically on mathematics teacher preparation for Black male mathematics learners. In her classes of "mostly White, female, middle-to-upper-class teachers," McGee encountered "fears and perceived inabilities to effectively teach… low-income, Black male students. They almost demanded that I provide them with a scripted blueprint for teaching Black males," she explained (Milner et al., 2013, p. 254).

Sheppard (2009) studied pre-service mathematics teachers' experiences mentoring and tutoring Black male students. He found that the pre-service teachers did not succumb to judgmental and condemnatory acts but were able to unearth the students' hidden or unmet mathematical potential. Sheppard also reported that providing pre-service teachers with an opportunity to get to know Black male students helped to inform their instruction. He stated that prospective mathematics teachers' field experiences should be situated in a context that provides them autonomy to adjust lessons to authentically integrate Black male students' experiences. Collectively, this body of scholarship speaks to the need to rethink mathematics teacher education programs for Black male students.

Rethinking Faculty and Pre-Service Teacher Preparation in Mathematics Education for Black Male Students

In Danny Martin's (2007) article *Beyond Missionaries or Cannibals: Who Should Teach Mathematics to African American Children?*, he critically analyzes and problematizes notions of "highly qualified mathematics teachers" and focuses on the question, Who should teach mathematics to African American children? He uses the term "missionaries" to connote "teachers who must save African American children from themselves and their culture" (p. 13). Martin uses the term "cannibals" to paint an extreme image of teachers who focus solely on mathematics content without any regard for students' lived experiences. He argues that the answer to his question largely revolves around,

> (a) The simplistic ways in which the aims and goals of mathematics education for African American (male) children are framed (i.e., closing the so-called achievement gap, increasing course enrollments, preparing students for the workforce) and (b) the problematic ways in which African American (male) children are socially constructed as learners with particular kinds of deficiencies in relation to students who are identified as white and Asian (male). (Martin, 2007, p. 14)

He calls for researchers, policy makers, and teacher educators to rethink what constitutes "highly qualified mathematics teachers in contexts predominated by African American (male) children" (Martin, 2007, p. 7). In this chapter, we respond to Martin's call by urging mathematics teacher educators to think about their role in preparing pre-service teachers to teach Black male students. We use Martin's (2007) conceptualization of highly qualified teachers for African American students as a guide to focus on Black males and rethink teacher education in mathematics. He states that teachers of African American students must

> (a) develop deep understanding of the social realities experienced by these students, (b) take seriously one's role in helping to shape the racial, academic, and mathematics identities of African American learners, (c) conceptualize mathematics not just as a school subject but as a means to empower African American students to address their social realities, and (d) become agents of change who challenge research and policy perspectives that construct African American children as less than ideal learners and in need of being saved or rescued from their blackness. (Martin, 2007, p. 25)

To build a cadre of teachers who conceptualize mathematics as described by Martin for Black male students, teacher education programs must rethink and restructure their programs and faculty. More specifically, mathematics teacher educators should utilize Black male students' lived experiences in the development of programs.

Recommendations for Mathematics Teacher Education Programs

Using the experience lens, we offer suggestions for teacher education programs to prepare pre-service teachers to teach Black male students. Mathematics teacher education programs must hire faculty members with critical consciousness about issues of race, racism, class, and gender as well as a sociopolitical perspective of the conditions of Black male students in society, schools, and mathematics spaces. Martin (2007) argues that mathematics "teacher educators cannot begin to adequately address the question of who should teach mathematics to [Black male students] without openly and honestly addressing issues of race, racialization, and identity" (p. 24). In many teacher preparation programs, curricular and programmatic priorities are determined by the typically White faculty members and administrators. These programs must also recruit Black male faculty members whose scholarship focuses on Black male students (Berry, 2008; Davis, 2014; Jett, 2013; Sheppard, 2009; Terry & McGee, 2012). These faculty members will be able to expose pre-service teachers to strength-based research surrounding high-achieving Black male students in mathematics education. This exposure will help challenge deficit beliefs and perspectives (Milner *et al.*, 2013; Sheppard, 2009). Mathematics teacher education programs must also center issues of race, racism, gender, classism, and other forms of oppression in the curriculum, practicum, and student teaching experiences (Martin, 2007, 2009; Sheppard, 2009). These critical issues will never find their way into the curriculum without the efforts of faculty members who possess a critical consciousness or sociopolitical perspective. Faculty members should ensure that the curriculum emphasizes social justice in the teaching and learning of mathematics (Gutstein, 2003, 2006; Larnell, Bullock, & Jett, 2016) as well as culturally relevant/responsive mathematics education (Greer, Mukhopadhyay, Powell, & Nelson-Barber, 2009), critical mathematics (Frankenstein, 2012), and ethnomathematics (d'Ambrosio, 1985;

Powell & Frankenstein, 1997) to challenge pre-service teachers' existing paradigms of the supposed neutrality and objectivity of mathematics as it is currently taught. This type of instruction will anchor teacher preparation in the sociopolitical context in which mathematics is taught.

To reach Black male students, pre-service mathematics teacher candidates must be culturally, socially, politically, and contextually relevant and social justice oriented. They must also have a deep knowledge of mathematics content to accompany their pedagogical approach. Their practicum and internship experiences should expose them to teachers who are culturally, socially, politically, and contextually relevant and social justice oriented. These experiences should prepare them to teach in diverse classrooms with students whose culture and lived experiences vary drastically from their own. Mathematics teacher educators must provide in-service teachers with ongoing professional development grounded in culturally, socially, politically, and contextually relevant and social justice oriented mathematics and pedagogical approaches.

Pre- and in-service teachers' conceptualization, thinking, and understanding of assessments must be expanded to align with culturally, socially, politically, and contextually relevant and social justice oriented mathematics and pedagogical approaches. These teachers must understand the history of standardized testing and its links to intelligence testing, eugenics, and scientific racism that was used to justify claims that Black people were intellectually inferior and mathematically inept (Davis & Martin, 2008). This process is needed to help teachers deconstruct their understanding of assessments as objective, unbiased, culture-free tools that provide an accurate portrayal of students' mathematical ability.

Additionally, mathematics teacher educators, and pre- and in-service teachers need a new lens that will help them recognize that standardized assessment results are a snapshot in time, not a finite calculation of ability. Moreover, assessments do not fully capture a students' capability or the experiences they may have had the night before or prior to the test. Many non-academic related factors can impact a student's performance on an assessment. To conceptualize assessments, Lee (1998) provides a culturally relevant lens that aligns with Martin's (2007) experience lens. Lee (1998) posits that Boykins' (1994) *Afro-cultural ethos* can serve as the foundation for culturally responsive, performance-based assessments that evidence vervistic performances, are socially situated, are multimodal and are rooted in community empowerment and transformation. Using an experience lens, teachers are better equipped to employ other methods beside achievement-based assessments to determine Black males' mathematical ability. Instead, they can use project and performance-based assessments that allow Black males to demonstrate not only their knowledge of the topic but also their ability to apply them to issues that impact their community and people.

Mathematics teacher education programs and faculty members must also provide all pre-service teachers with opportunities to spend time with Black males and their families to get to know them in non-academic spaces. These types of opportunities should be integrated into all aspects of pre-service teacher preparation. Pre- and in-service mathematics teachers must learn to see the humanity in Black male students and understand that they are not a monolithic group. In our experiences, all Black male students have been expected to think and behave in stereotypic ways. Variations in experiences have often been dismissed and ignored in discussions of teacher preparation.

Mathematics teacher education programs must prepare pre-service teachers with knowledge, skills, and dispositions to positively shape Black male students' racial and mathematics identities. Program faculty members must also actively recruit pre-service Black male teachers into the program and provide them with the proper support structures. To accomplish this, teacher education programs should partner with "grow your own teacher" programs that seek to work with students as young as middle school to spark their interest in teaching. For instance, in their study examining the experiences of 22 Black male 11th and 12th grade students who participated in Pathways2Teaching (P2T), a "grow your own teacher" program, Goings and Bianco (2016) found that Black males were inspired to enter the teaching profession when they had the chance to work with students in elementary schools. These experiences allowed them to apply what they learned in their P2T class, which focused on urban education. These types of teaching experiences are critical to developing a mathematics teacher workforce that is prepared and has the will to impact the lives of their Black male students positively.

Conclusion

As scholars committed to advancing our knowledge base on Black male students and teachers, we are concerned that there has not been a laser-like focus on the preparation of mathematics teachers working with Black males. As noted throughout the chapter, we must have a sense of urgency on this issue. The change will not come by happenstance; it must be deliberate and direct. It will take the dedicated work of mathematics teacher educators and the programs they produce to ensure that mathematics teachers enter the classroom with the experience, expertise, and perspectives necessary to effect positive change and elicit the mathematics ability in Black male students. Mathematics teachers must be prepared to face the intersection of race, racism, class, gender, culture, and mathematics, all while creating a welcoming learning environment. Mathematics teacher educators must ensure that their curricular offerings provide more than a single diversity course. By the time they enter the classroom, new mathematics teachers must be well acquainted with issues of race, power, and privilege, and understand how to support the mathematical learning of Black males. Mathematics should be seen not as mere numbers and equations, but as a tool to disrupt and dismantle systemic inequities. This understanding will bring about lasting change.

References

Adler, J., Ball, D., Krainer, K., Lin, F. L., & Novotna, J. (2005). Reflections on an emerging field: Researching mathematics teacher education. *Educational Studies in Mathematics, 60*(3), 359–381.

Berry III, R. Q. (2005). Voices of success: Descriptive portraits of two successful African American male middle school mathematics students. *Journal of African American Studies, 8*(4), 46–62.

Berry III, R. Q. (2008). Access to upper-level mathematics: The stories of successful African American middle school boys. *Journal for Research in Mathematics Education, 39*(5), 464–488.

Berry III, R. Q., & McClain, O. L. (2009). Voices, power, and multiple identities: African-American boys and mathematics success. *New England Mathematics Journal, 41*, 17–26.

Berry III, R. Q., Thunder, K., & McClain, O.L. (2011). Counter narratives: Examining the mathematics and racial identities of Black boys who are successful with school mathematics. *Journal of African-American Males in Education, 2*(1), 10–23.

Boykin, A. W. (1994). Afrocultural expression and its implications for schooling. In E. R. Hollins, J. E. King, & W. C. Hayman (Eds.), *Teaching diverse populations: Formulating a knowledge base* (pp. 50–51). Albany, NY: State University of New York.

Brown, I. A., Davis, T. J., & Kulm, G. (2011). Pre-service teachers' knowledge for teaching algebra for equity in the middle grades: A preliminary report. *The Journal of Negro Education, 80*(3), 266–283.

Corey, D. L., & Bower, B. L. (2005). The experiences of an African American male learning mathematics in the traditional and the online classroom—A case study. *The Journal of Negro Education, 74*(4), 321–331.

d'Ambrosio, U. (1985). Ethnomathematics and its place in the history and pedagogy of mathematics. *For the Learning of Mathematics, 5*(1), 44–48.

Davis, J. (2014). The mathematical experiences of Black males in a predominantly Black urban middle school and community. *International Journal of Education in Mathematics, Science and Technology, 2*(3), 206–222.

Davis, J., & Martin, D. B. (2008). Racism, assessment, and instructional practice: Implications for mathematics teachers of African American students. *Journal of Urban Mathematics Education, 1*(1), 10–34.

Frankenstein, M. (2012). Beyond math content and process: Proposals for underlying aspects of social justice education. In A. Wager and D.W. Stinson, *Teaching mathematics for social justice: Conversations with educators*, 49–62. Reston, VA: National Council of Teachers of Mathematics

Goings, R. B., & Bianco, M. (2016). It's hard to be who you don't see: An exploration of Black male high school students' perspectives on becoming teachers. *The Urban Review, 48*, 628–646.

Goings, R. B., & Ford, D. Y. (2018). Investigating the intersection of poverty and race in gifted education journals: A 15-year analysis. *Gifted Child Quarterly, 62*(1), 25–36.

Greer, B., Mukhopadhyay, S., Powell, A. B., & Nelson-Barber, S. (Eds.). (2009). *Culturally responsive mathematics education*. New York, NY: Routledge.

Gutstein, E. (2003). Teaching and learning mathematics for social justice in an urban, Latino school. *Journal for Research in Mathematics Education, 34*(1), 37–73.

Gutstein, E. (2006). *Reading and writing the world with mathematics: Toward a pedagogy for social justice*. New York, NY: Routledge.

Howard, T. C. (2013). *Black male(d): Peril and promise in the education of African American males*. New York, NY: Teachers College Press.

Howard, T. C., Flennaugh, T. K., & Terry, C. L., Sr. (2012). Black males, social imagery, and disruption of pathological identities: Implications for research and teaching. *Educational Foundations, 26*(1–2), 85–102.

Hrabowski, F., Maton, K., & Grief, G. (1998). *Beating the odds: Raising academically successful African-American males.* New York, NY: Oxford University Press.

Jett, C. C. (2009). *African American men and college mathematics: Gaining access and attaining success* (Doctoral dissertation). Athens, GA: University of Georgia.

Jett, C. C. (2011). "I once was lost, but now am found" The mathematics journey of an African American male mathematics doctoral student. *Journal of Black Studies, 42*(7), 1125–1147.

Jett, C. C. (2013). HBCUs propel African American male mathematics majors. *Journal of African American Studies, 17*(2), 189–205.

Jett, C. C. (2016). Building on our mathematical legacy of brilliance: A critical race reflective narrative. In B. L. McGowan, R.T. Palmer, J. L. Wood, & D. Hibbler (Eds.), *Black men in the academy: Narratives of resiliency, achievement, and success* (pp. 77–88). New York, NY: Palgrave Macmillan.

Ladson-Billings, G., & Tate, W. (1995). Towards a critical race theory of education. *Teachers College Record, 97*(1), 47–68.

Larnell, G. V., Bullock, E. C., & Jett, C. C. (2016). Rethinking teaching and learning mathematics for social justice from a critical race perspective. *Journal of Education, 196*(1), 19–29.

Lattimore, R. (2003). African-American students struggle on Ohio's high-stakes test. *Western Journal of Black Studies, 27*(2), 118.

Lattimore, R. (2005a). Harnessing and channeling African American children's energy in the mathematics classroom. *Journal of Black Studies, 35*(3), 267–283.

Lattimore, R. (2005b). African American students' perceptions of their preparation for a high-stakes mathematics test. *Negro Educational Review, 56*(2/3), 135.

Lee, C. D. (1998). Culturally responsive pedagogy and performance-based assessment. *Journal of Negro Education, 67*(3), 268–279.

Martin, D. B. (2007). Beyond missionaries or cannibals: Who should teach mathematics to African American children? *The High School Journal, 91*(1), 6–28.

Martin, D. B. (2009a). Researching race in mathematics education. *Teachers College Record, 111,* 295–338.

Martin, D. B. (Ed.). (2009b). *Mathematics teaching, learning and liberation in the lives of Black children.* New York, NY: Routledge.

Martin, D. B., & McGee, E. (2009). Mathematics literacy for liberation: Reframing mathematics education for African American children. In B. Greer, S. Mukhopadhyay, A. Powell, and S. Nelson-Barber, *Culturally responsive mathematics education* (pp. 207–238). New York, NY: Routledge

McGee, E. O. (2013). Threatened and placed at risk: High achieving African American males in urban high schools. *The Urban Review, 45*(4), 448–471.

McGee, E. O. (2014). When it comes to the mathematics experiences of Black pre-service teachers. Race matters. *Teachers College Record, 116*(6), 1–50.

McGee, E. O., & Martin, D. B. (2011). From the hood to being hooded: A case study of a Black male PhD. *Journal of African-American Males in Education, 2*(1), 46–65.

McGee, E. O., & Pearman, F. A. (2014). Risk and protective factors in mathematically talented Black male students: Snapshots from kindergarten through eighth grade. *Urban Education, 49*(4), 363–393.

McGlamery, S., & Mitchell, C. T. (2000). Recruitment and retention of African-American males in high school mathematics. *Journal of African-American Men, 4*(4), 73–87.

Milner, H. R., Pabon, A., Woodson, A., & McGee, E. (2013). Teacher education and Black male students in the United States. *REMIE Multidisciplinary Journal of Educational Research, 3*(3), 235–263.

Nasir, N. S., & Hand, V. (2008). From the court to the classroom: Opportunities for engagement, learning and identity in basketball and classroom mathematics. *Journal of the Learning Sciences, 17*(2), 143–179.

Noble, R. (2011). Mathematics self-efficacy and African American male students: An examination of models of success. *Journal of African American Males in Education, 2*(2), 188–213.

Noguera, P. A. (2009). *The trouble with black boys:... And other reflections on race, equity, and the future of public education.* Hoboken, NJ: John Wiley & Sons.

Oakes, J. (2005). *Keeping track: How schools structure inequality.* New Haven, CT: Yale University Press.

Pitts Bannister, V. R., Davis, J., Mutegi, J., Thompson, L., & Lewis, D. (2017). "Returning to the root" of the problem: Improving the social condition of African Americans through science and mathematics education. *In Catalyst: A Social Justice Forum, 7*(1), 2, 4–14.

Polite, V. C. (1999). Combating educational neglect in suburbia: African American males and mathematics. In V. C. Polite & J. E. Davis (Eds.), *African American males in school and society: Practices and policies for effective education.* New York, NY: Teachers College Press.

Polite, V. C., & Davis, J. E. (Eds.). (1999). *African American males in school and society: Practices and policies for effective education.* New York, NY: Teachers College Press.

Powell, A. B., & Frankenstein, M. (Eds.). (1997). *Ethnomathematics: Challenging eurocentrism in mathematics education.* Albany, NY: State University of New York Press.

Sealey-Ruiz, Y., & Lewis, C. W. (2011). Guest editorial: Transforming the field of education to serve the needs of the Black community: Implications for critical stakeholders. *The Journal of Negro Education, 80*(3), 187–190.

Sealey-Ruiz, Y., Lewis, C. W., & Toldson, I. (Eds.). (2014). *Teacher education and black communities: Implications for access, equity and achievement.* Charlotte, NC: Information Age Publishing.

Sheppard, P. A. (2009). Prospective teachers' experiences teaching mathematics to African American males. *Education, 130*(2), 226–231.

Sheppard, P. A. (2011). Experience-centered instruction as a catalyst for teaching mathematics effectively to African American students. *The Journal of Negro Education, 80*(3), 254–265.

Smith, J. P., & Cage, B. N. (2000). The effects of chess instruction on the mathematics achievements of southern, rural, Black secondary students. *Research in the Schools, 7*(1), 19–26.

Stinson, D. W. (2006). African American male adolescents, schooling (and mathematics): Deficiency, rejection, and achievement. *Review of Educational Research, 76*(4), 477–506.

Stinson, D. W. (2008). Negotiating sociocultural discourses: The counter-storytelling of academically (and mathematically) successful African American male students. *American Educational Research Journal, 45*(4), 975–1010.

Stinson, D. W. (2011). When the "burden of acting White" is not a burden: School success and African American male students. *The Urban Review, 43*(1), 43–65.

Stinson, D. W. (2013). Negotiating the "White male myth": African American male students and success in school mathematics. *Journal for Research in Mathematics Education, 41*, 1–31.

Stinson, D. W., Jett, C. C., & Williams, B. A. (2013). Counterstories from mathematically successful African American male students: Implications for mathematics teachers and teacher educators. In J. Leonard & D. B. Martin (Eds.), The *brilliance of Black children in mathematics: Beyond the numbers and toward new discourse* (pp. 221–245). Charlotte, NC: Information Age Publication.

Terry Sr., C. L. (2010). Prisons, pipelines, and the president: Developing critical math literacy through participatory action research. *Journal of African American Males in Education, 1*(2), 73–104.

Terry Sr., C. L. (2011). Mathematical counter story and African American male students: Urban mathematics education from a critical race theory perspective. *Journal of Urban Mathematics Education, 4*(1), 23–49.

Terry Sr., C. L., & McGee, E. O. (2012). "I've come too far, I've worked too hard": Reinforcement of support structures among Black male mathematics students. *Journal of Mathematics Education at Teachers College, 3*(2), 73–85.

Thompson, L., & Davis, J. (2013). The meaning high-achieving African-American males in an urban high school ascribe to mathematics. *The Urban Review, 45*, 490–517.

Thompson, L. R., & Lewis, B. F. (2005). Shooting for the stars: A case study of the mathematics achievement and career attainment of an African-American male high-school student. *High School Journal, 88*(4), 6–18.

Toward a Framework for Culturally Relevant Inquiry-Based Science Pedagogy

VANESSA DODO SERIKI

Abstract

Culturally relevant pedagogy (CRP), as posited by Ladson-Billings (2009) and Inquiry-based science instruction (Marshall et al., 2010) are two pedagogical approaches that have, separately, been shown to positively influence achievement among students who experience it (Gutstein, Lipman, Hernandez, & de los Reyes, 1997; Lopez, 2011; & Marx, Blumenfeld, Krajcik, Fishman, Soloway, Geier, & Tal, 2004). In this chapter, culturally relevant pedagogy and inquiry-based science instruction are juxtaposed to highlight the overlap and distinctive features of each, illustrate why it is crucial for science teachers of African American students to have and use CRP and Inquiry-Based Science Instruction (IbSI) in their daily practice, and to advance a science pedagogical framework that combines the two. Such a framework, which is tentatively identified as Culturally Relevant Inquiry-based Science Pedagogy (CRISP), can be used to aid both preservice and in-service science teachers in creating equitable science classrooms and learning opportunities that allow learners—particularly African American children—to access and experience high-quality science instruction that is student-centered and culturally relevant. In addition to unpacking and defining the features of CRISP, this chapter opens with a short vignette that is demonstrative of the importance of bringing the two approaches together.

Opening Vignette

The day started like any other, and I arrive at Watershed High School ready to observe a science lesson, which was written by a teacher candidate who had gone out of his way to integrate students' funds of knowledge and cultural capital into the lesson. Kevin Myer, a White physics teacher candidate from a rural county in a Midwestern state, was nervous about his plan but was confident that the students would engage and learn as he had gone through great lengths to ensure that the lesson would be interesting and relevant to the students. The students were high school juniors attending a large high school in a major metropolitan city in the Midwestern United States. The lesson, which was on sound and wavelengths, involved the use of music to help make the lesson meaningful to them. As he developed the lesson, he had asked students about their favorite music and selected from this list various songs that the student groups would use to collect data on sound levels and sound waves using a Venier™ sound meter and microphone. Eager to see the student-centered lesson, I arrived a bit early and was quickly unsettled as Kevin was waiting for me in the office. He was extremely anxious and quite frustrated because at the last minute his mentor teacher instructed him to change his lesson by removing the music and merely providing the students with notes.

Recognizing our precarious position as guests in the mentor's classroom, I calmed Kevin and instructed him to follow the teacher's plan. After class, I planned to engage Kevin in a post-observation debriefing, but before we could get started his mentor wanted to express his frustration with Kevin's attempt to infuse music into the lesson. He told us, "This is physics class and the students must learn the content. We don't have time for this culture stuff. Their music and culture has nothing to do with science!" Taken aback by his comment, I politely responded, "You don't have time *not* to place students at the center of their learning. When students can make connections between the content and aspects of their lives, learning will become more meaningful." To which he responded, "Yeah? Well, I don't have time to learn about all the students' likes and dislikes, they will just have to study."

Introduction

The mentor's stance on integrating students' culture into teaching is a position that is not uncommon. It is riddled with assumptions and misconception about what it means to teach in culturally relevant or responsive ways (Gay, 2010; Ladson-Billings, 2009). His stance is so prevalent that science education scholars, who ascribe to CRP, often discuss how to counter such sentiment that argues that science is culture free. My response to such pushback is that science is not culture free. Human beings engage in science—all types of science whether Western science standards recognize it or not. Humans have culture, and they bring their cultural ways of knowing, understanding, and interacting with the world to science. In fact, not only is science a human endeavor, which intricately links science and culture, it is also not apolitical nor completely objective. How we understand science is based on how we have been taught science. Traditionally science is taught as the absolute truth despite the nature of science describing science as a human endeavor that is based on the knowledge we have available at that time—thus the tentative nature of science.

In this chapter, I critique the underlying assumptions and misconceptions associated with the pushback, using Jegede and Aikenhead's conception of border crossing and Western science (Aikenhead, 1996; Jegede & Aikenhead, 1999), as a framework. I highlight the parallels between Culturally Relevant Pedagogy (CRP) and Inquiry-based Science Instruction (IbSI) to merge the two models and advocate for the merger of these two because such a merger would allow teachers and researchers to confront their dispositions (beliefs) and interrogate their classroom practices. The significance of conceptualizing this model has implications for all students. However, merging CRP and IbSI has more significant implications for the teaching and learning of science to and by African American students. The implications are most important as African American students continue to be underrepresented and underperform as compared to their non-African American peers in STEM fields in general and science in particular. While there is a myriad of reasons for the underrepresentation and underperformance, one glaring explanation is the way in which they are granted, or not granted, access to science and STEM learning opportunities within K–12 classrooms.

For many African American students, access to quality science learning experiences is lacking as some teachers, as Prime and Miranda (2006) found, do not believe African American students possess "… the high intellectual and attitudinal demands (p. 526)" needed for science. As a result, teachers made curricular decisions that changed the quality and quantity of science content to which these students were exposed (Prime & Miranda, 2006). Essentially, their study made explicit the link between teachers' beliefs and their subsequent action in the classroom. Moreover, the work of Bryan and Atwater (2002) indicated that "… beliefs are part of a group of constructs that describe the structure and content of a person's thinking that is presumed to drive his/her actions" (p. 823). Thus, when teachers believe their African American students lack the intellectual capability and the requisite knowledge needed to engage in authentic science, then teachers will make decisions and behave in ways that limit students' access to relevant, meaningful, and rigorous science learning opportunities.

Countering the Pushback: Western Science and Border Crossing

The notion that science is culture- and value-free is not new. In Aikenhead's (1996) description of Western culture and school science, he explains where this idea likely—yet mistakenly—originated. He posited, "unfortunately, the 'taught' science curriculum, more often than not, provides students with a stereotypic image of science: socially sterile, authoritarian, non-humanistic, positivistic, and absolute truth" (Aikenhead, 1996, p. 10). This presentation of science has led some teachers to believe that science stands apart from, and not within, culture, and thus is culture-free. Therefore, attention to culture is unnecessary in the teaching and learning of science. Aikenhead (1996) refutes the idea of science being culture-free as he understands that science is situated within Western culture, which has a shared system of meaning and symbols. This shared system, which is based on "white male middle-class Western system[s]" (Aikenhead, 1996, p. 8) of thinking and viewing the world influences those other cultures (i.e., science) that are situated within them. Thus, the shared system of Western culture is also shared by the culture of science thereby yielding Western science.

Students who are accustomed to the shared systems within Western culture would find greater ease in navigating between their home culture, which is akin to Western culture, and the culture of Western science, than those who have a different home culture. Jegede and Aikenhead (1999) identified and labeled these students as "potential scientists" and "other smart kids" (p. 50).

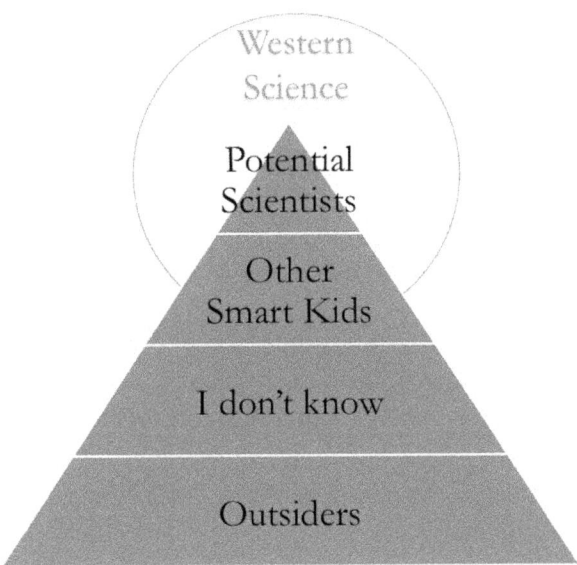

Figure 6.1: *Types of Border Crossings* (adapted from writings of Jegede and Aikenhead, 1999; figure representation is author's own.)

The clashing of cultures challenges the remaining students who are not able to easily navigate or reconcile the discrepancy between their home culture and the culture of Western science. These students fit into one of the two remaining categories shown in Figure 6.1 (Jegede & Aikenhead, 1999). These categories are based on the students' ability to navigate between their home culture and the culture of Western science. As depicted in Figure 6.1, and in the Jegede and Aikenhead (1999) article, only the top two levels of students (potential scientists and other smart kids) can easily navigate Western Science. That does not imply that the other classifications of students cannot access science, it just implies that their navigation is fraught with challenges that are sometimes difficult to overcome.

The mentor teacher in the vignette mistakenly believes science is culture free, and his instruction appeals only to those students Jegede and Aikenhead (1999) identified as "potential scientists" and "other smart kids" (p. 50). Given that the home culture of African American students may differ from that of Western culture and Western science—these students are likely to struggle in classrooms where the teacher rejects the significance and impact of culture on science teaching and learning. I posit that science education—particularly for African American children—needs three things, which are:

1. Teachers who function as cultural workers (as described by Freire, 1998);
2. Exposure to CRP (as posited by Ladson-Billings, 2009); and
3. Experience with quality IbSI (Marshall, Horton, & White, 2009).

Moreover, I argue that the melding of points two and three (CRP and IbSI) generates a framework by which science teachers can self-monitor their practices and growth to create more equitable science classrooms wherein African American children can thrive.

Teachers as Cultural Workers

Teachers who are cultural workers understand that education, as Freire (1998) asserted, is not a neutral process—teaching is a political act. This idea makes teachers political agents who must possess particular qualities. These qualities include humility, lovingness, courage, tolerance, decisiveness, patience, and impatience (Freire, 1998). All of these qualities were expertly defined and illustrated in Freire's fourth letter to teachers (1998). While understanding these are important, what is most striking is that teachers who possess these qualities are continually seeking justice by challenging the status quo, which requires courage; courage to utilize approaches to instruction that are innovative and student-centered. The positionality of African American children with respect to Western science as a culture that is alien to their home cultures demands that science teachers of African American children must be culture workers.

Culturally Relevant Pedagogy

CRP is a pedagogical model that functions to empower students to achieve, academically, with their culture as the core basis of learning and to provide opportunities to apply their learning to challenge the current social order (Ladson-Billings, 2009). Although CRP is nestled within multicultural literature and prominently found in literacy and social studies education, it is relevant and applicable across all subject areas. Given the continued and persistent underrepresentation of African American students in science, I argue that it is an appropriate and necessary model for science teaching and learning.

As a pedagogical model used most often with P–12 students, CRP emerged from the work of Ladson-Billings (2009). Her foundational study, *The Dreamkeepers: Successful teachers of African American children*, presents CRP as a model for increasing achievement using students' culture while empowering them to critically examine the world using newly acquired knowledge. This approach is similar to that of critical pedagogy (Ladson-Billings, 1995) because it functions to empower all students who experience it. CRP rests upon three propositions:

1. Students must experience academic success (Academic Achievement);
2. Students must develop cultural competence (Cultural Competence); and
3. Students must develop a critical consciousness (Socio-political Consciousness) through which they challenge the status quo of the current social order. (Ladson-Billings, 1995, p. 160)

Within these three tenets Ladson-Billings (2001) identified behaviors and understandings indicative of culturally relevant teachers. The following three lists detail these specific teacher actions:

Academic Achievement
- The teacher presumes that all students are capable of being educated;
- The teacher delineates what achievement means in the context of his or her classroom;
- The teacher knows the content, the learner, and how to teach content to the learner;
- The teacher supports a critical consciousness toward the curriculum; and
- The teacher encourages academic achievement as a complex conception not amenable to a single, static measurement. (Ladson-Billings, 2001, p. 74)

Cultural Competence
- The teacher understands culture and its role in education;
- The teacher takes responsibility for learning about students' culture and community;
- The teacher uses student culture as a basis for learning; and
- The teacher promotes a flexible use of students' local and global culture. (Ladson-Billings, 2001, p. 98)

Sociopolitical Consciousness
- The teacher knows the broader sociopolitical context of the school community, nation, and world;
- The teacher has an investment in the public good;
- The teacher plans and implements academic experiences that connect students to the broader social context; and
- The teacher believes that students' success has consequences for his or her quality of life. (Ladson-Billings, 2001, pp. 120–121)

In addition to the three propositions and indicators, Ladson-Billings (1990) in her early analysis of successful teachers of Black students, identified three themes that related to teachers' beliefs about their success with Black students. These themes were:

1. Conception of self and others;
2. Social relations; and
3. Conceptions of knowledge (Ladson-Billings, 1990).

In her explication of these teacher-held views she described how they translate into teacher beliefs and behaviors within the classroom. For instance, culturally relevant teachers' conceptions of themselves and others allow them to "see [themselves] as artists, part of the community to which they are giving back, and they help students make connections between their community, national, ethnic, and global identities." (Ladson-Billings, 1990, p. 340). She contrasted these views with those held by assimilationist teachers. These conceptions, although not specific behaviors, provide insight into why culturally relevant teachers utilize particular instructional practices—if a teacher views their role as "pulling knowledge out" then they are less likely to use teacher-directed instruction (Ware, 2006). Instead, they would engage students in active learning (Prince, 2004) that connected what they were learning to their various identities (Ladson-Billings, 1990).

Ladson-Billings (1990) notes that the social relations that culturally relevant teachers have with their students are essential. Such teachers have fluid relationships with their students that extend "beyond the classroom into the community" (p. 340). In such a classroom collaboration is valued and welcomed—it is not just a place where students are, but rather a space used by a community of learners. Parsons, Travis, and Smith-Simpson (2005) substantiated the importance of relationships in their study of culturally congruent science instruction in 8th-grade science classrooms. The fluidity of relationships—wherein hierarchical relationships between the teacher and students are nonexistent—allow the interactions between the teacher and students to be in constant flux where positions are negotiated and renegotiated throughout the exchange. Ladson-Billings (1990) further posited that students learning together and being responsible for one another exemplify this dynamic understanding of relationships.

> **Inquiry** is a multifaceted activity that involves making observations; posing questions; examining books and other sources of information to see what is already known; planning investigations; reviewing what is already know in light of experimental evidence; using tools to gather, analyze, and interpret data; proposing answers, explanations and predictions; and communicating the results. Inquiry requires identification of assumptions, use of critical and logical thinking, and consideration of alternative explanations (NRC, 1996, p. 23)

Figure 6.2: Definition of Inquiry (figure is author's own).

The last yet significant theme, from Ladson-Billings' (1990) early work, is the teacher's conception of knowledge. Teacher's view of knowledge impacts the type of learning opportunities they extend to their students. Culturally relevant teachers believe that knowledge is dynamic—it is continually being recreated, recycled, and shared within the learning community (Ladson-Billings, 1990, p. 340). Rather than lamenting about the requisite knowledge students lack, they help them develop the knowledge and skills they need to be successful. Also, these teachers teach students how to view knowledge critically.

Inquiry-Based Science Instruction

IbSI is predicated on constructivist learning theories. These theories of cognition posit that learners do not enter the learning space as blank slates, instead, they bring a host of experiences, knowledge, and cultural referents that inform what they know and how they came to know it. Although we cannot look to constructivism to explain or substantiate specific teaching practices, it is useful in explaining and examining how students engage in learning—or knowledge construction. However, based on constructivist theories of learning, teachers can implement instructional practices that support students who learn in this way. Several scholars have examined these types of practices (i.e., Marshall et al., 2009; Sawada et al., 2002), and have developed observation protocols to measure the level of inquiry that occurs in classrooms. While there are many tools, such as observation protocols, that measure inquiry and offer valuable data about it, none of them served as a tool to guide teachers through a reflective process in which goals on the quantity and quality of inquiry being used in their classroom could be set. However, the development of the Electronic Quality of Inquiry Protocol (EQUIP) was constructed based on the National Science Education Standards (NSES) (National Research Council [NRC], 1996) definition of inquiry (Figure 6.2) and it clearly delineates observable instructional practices that are characteristic of inquiry-based instruction. The NSES are no longer in use, but the processes, skills, and practices associated with inquiry are still relevant and quite prominent in the current Next Generation Science Standards (NGSS), and inquiry-based instruction remains a crucial feature of science instruction.

IbSI, in general, requires teachers to adopt a constructivist view of learning. The EQUIP (Marshall et al., 2009) outlines and describes inquiry-based instruction, and it measures the quantity and quality of inquiry occurring in classrooms—particularly science and mathematics classrooms. The EQUIP is divided into four areas: Instruction, Discourse, Assessment, and Curriculum, each of which contains specific areas of focus. The instrument is a descriptive rubric that provides a benchmark for teacher performance. In many ways, this instrument provides explicit criteria that clearly define inquiry-based instruction.

In using the instrument, Marshall et al. (2009) positioned level three, *proficient inquiry,* as the target benchmark, to which science and mathematics teachers should aspire. The descriptive nature of the instrument allows practitioners to engage in self-reflection and goal setting for achieving the target benchmark. Proficient inquiry instruction is distinguished by students having opportunities to engage in and with content before having the content explained to them (Marshall, Smart, & Alston, 2017). IbSI explicitly addresses instructional practices and requires teachers to have a specific disposition toward teaching and learning that allows them to construct learning opportunities and experiences that are meaningful to students. These learning opportunities are student-centered and afford students the chance to explore content before being formally introduced to concepts. Inquiry—as described by Marshall et al. (2010)—requires content depth on the part of teachers that would enable them to connect content to the bigger picture. While a teacher could visualize the bigger picture as being connected to the local, global, social, political, economic, historical, and cultural context, IbSI does not require these particular connections. CRP, on the other hand explicitly addresses teachers' disposition toward teaching and learning. These dispositions are essential for all teachers and rest upon the tenets of CRP and what teachers value and believe about knowledge, social relations, and self and others. While IbSI offers specific practices, CRP influences the practices teacher chose to use. Moreover, CRP requires the content be used to address or redress issues within students' local and global social, political, economic, and cultural context.

Juxtaposition of CRP and IbSI

At first glance, CRP and IbSI seem different with one being more appropriate and aligned with contemporary approaches to science instruction and the other being appropriate for general or non-content-specific instructional practices and beliefs. Despite the distinctive features, I offer four assertions about the relationship between the two:

1. Both, CRP and IbSI, are appropriate for all content areas;
2. One approach, CRP, also functions as a set of pedagogical dispositions that supports IbSI;

3. To engage children, particularly African American children, in meaningful science learning—teachers must adopt a CRP disposition to fully engage IbSI; and
4. The tenets and indicators of CRP can be found throughout the four constructs of IbSI.

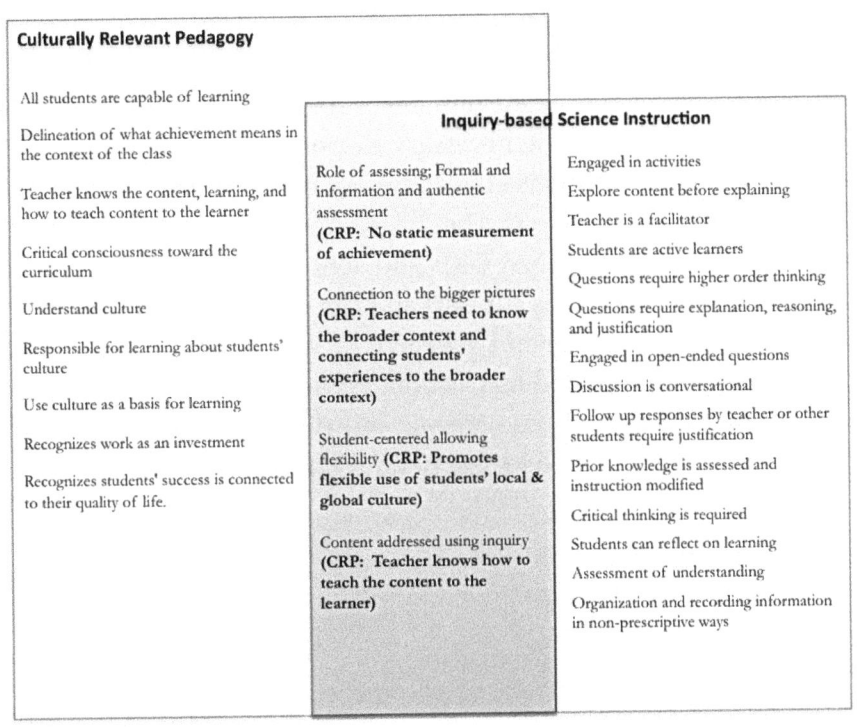

Figure 6.3: Relationship Between CRP and IbSI.

Initial thinking about the alignment between CRP and IbSI was based on the assumption that both exist as pedagogical models or approaches that have been shown to positively influence learning both in STEM-based courses and other content areas. Indeed, there are studies supporting this assumption (for example, see Aronson & Laughter, 2016; Kahle, Meece, & Scantlebury, 2000). However, together with the alignment and distinctive features (Figure 6.3), these two models can change the status quo for African American students in science. Ladson-Billings (2006) encouraged educators to move from "doing" CRP to "being" culturally relevant. Being culturally relevant requires that teachers possess a particular disposition toward their students that is rooted in what they value and believe about the conception of knowledge, social relations, and conceptions of themselves and others (Ladson-Billings, 1990). Teachers who possess such a disposition would, without question, engage students in IbSI.

Unpacking the Alignment

The relationship between CRP and IbSI is such that to do one well (IbSI), a teacher must assume the necessary dispositions of the other (CRP). Tenets of CRP can be observed throughout the inquiry-based practices of teachers who ascribe to it. The diagram in Figure 6.4 illustrates the nested relationship between CRP and IbSI. The figure suggests that CRP and its various tenets support and can be found throughout IbSI. However, there is alignment between CRP and IbSI in the areas of assessment, curriculum, and instruction. While none of these categories, curriculum, instruction, assessment, and discourse, serves as a hallmark of IbSI, they are all necessary and required features of reaching.

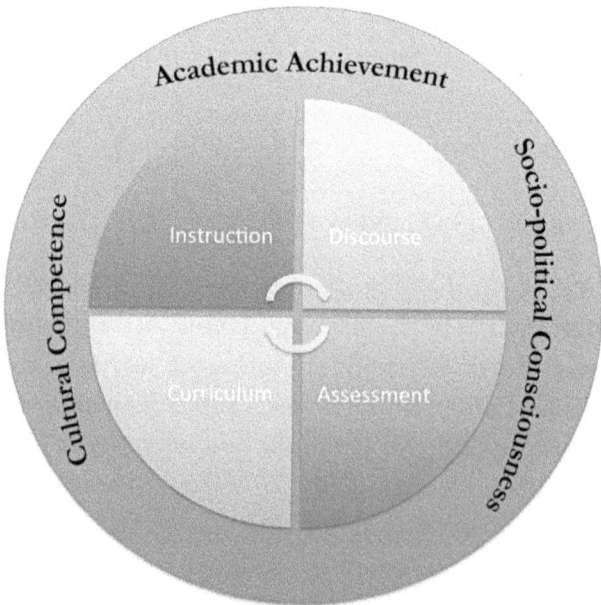

Figure 6.4: Nested Relationship Between CRP and IbSI.

Assessment. Assessment is a vital part of instruction as it provides the teacher with important information about the effectiveness of the curriculum, instruction (Popham, 2017), and students' prior knowledge. In inquiry-based instruction, assessment serves the primary function of providing data to the teacher that can be used to modify or adjust instruction. CRP encourages, just as IbSI does, teachers to use varied measurements (assessments) to monitor student achievement. While CRP does not call for modification to instruction, culturally relevant teachers understand that such adjustment is necessary to ensure that students experience success (Ladson-Billings, 2009). Moreover, assessment can, and should, be used during instruction as a learning tool. Proficient inquiry necessitates that assessment be process-focused and utilizes critical thinking in order for students to develop conceptual understanding (Marshall *et al.*, 2010). Also, crucial for student learning is creating opportunities for students to reflect on their learning in ways that allow them to make sense of the concepts being taught. Together CRP and IbSI are dependent on the use of authentic assessment measures. While CRP does not explicitly address formal and informal assessment, it most certainly critiques the notion of using single static measures of student success.

Curriculum. The alignment between CRP and IbSI from the perspective of curriculum is quite robust. Both require a student-centered focus. However, CRP calls for teachers to place students at the center of instruction by using their culture as the basis of learning (Ladson-Billings, 2009). IbSI advocates for students having opportunities, during instruction, to design their own explorations (Marshall *et al.*, 2010). Merging the two yields a focus on student centrality that allows students to design explorations that are based on their frames of reference (Gay, 2010) and culture (Ladson-Billings, 2009). As with any good pedagogical model, both CRP and IbSI require teachers to make effective use of content standards. IbSI addresses the use of science content standards and CRP necessitates that teachers know their content and how to teach the content to students (Ladson-Billings, 2009). IbSI does not address the content standards, explicitly, in the same way but proficient inquiry promotes the use of inquiry to teach the [content] standards. It provides the specificity to knowing the content standards and how to teach the content to students (i.e., inquiry). In regard to content depth of the curriculum, proficient inquiry involves "... significant connections to the bigger picture" (Marshall et al., 2010). This significant connection, within CRP, includes local and global connections that are contextually situated (i.e., context refers to the bigger picture of students' social, historical, economical, and cultural realities).

Instruction. The alignment between CRP and IbSI in the area of instruction is not as evident as it is in other areas. The reason for this reflects IbSI's focus on practice and CRP's focus on dispositions. The place where the two models intersect is in the role of the teacher and students and instructional strategies. While neither provides specific or explicit instructional strategies, they both advocate for the use of strategies that cultivate active learning (Ladson-Billings, 2001; Marshall et al., 2010; Prince, 2004). Concerning the teacher and students' role, both call for teachers to function as facilitators or coaches. Students are active participants in the learning process. Another area of alignment that is not readily evident involves what teachers believe about the conception of knowledge. Culturally relevant teachers recognize that knowledge is socially constructed and is recycled continuously and reconstructed (Ladson-Billings, 1990). Possessing this disposition pushes teachers to do what IbSI encourages, which is to create learning experiences that allow students to apply concepts rather than passively acquiring the concepts.

Revealing and Using the Framework

To date, there is no observation protocol in use for CRP like there are for IbSI. However, the indicators that Ladson-Billings (2001) laid out could conceivably be transformed into an analytic rubric for use in any classroom. For this conception of merging CRP and IbSI, CRP is best measured through the implementation of a rubric that would allow those striving to be culturally relevant to self-evaluate, set goals, and continue moving toward "being" culturally relevant (Ladson-Billings, 2006). A merged framework would highlight—precisely—the alignment between the two while also retaining the distinctive features of each. Given the recommendation of using an analytic rubric, the protocol should contain criteria (i.e., indicators of both CRP and IbSI) as well as levels that appropriately describe the observation of both in practice. The use of such an observation protocol or tool is ideal for these constructs for several reasons. They include the following:

1. The tool would identify target practices for which teachers could strive;
2. Teachers could self-assess and set goals; and
3. It offers flexibility—to the user—as it does not require that the unit of analysis be a particular set of strategies but can utilize practices that are appropriate to their given situation.

Culturally Relevant Inquiry-Based Science Pedagogy (CRISP). The framework being conceptualized is called Culturally Relevant Inquiry-based Science Pedagogy—CRISP for short. This initial conceptualization does not remove any tenets or indicators from either model, but rather merges those aspects that are similar and includes the distinctive features of each. For example, when looking for ways the teacher connects the content to the bigger picture, the CRISP model would encourage connections between the content and the students' local, global, and national community and identities.

As the framework is transformed into a functional tool, it will utilize a developmental design supported by an analytic observation tool that places teacher practices on a continuum (i.e., No culturally relevant inquiry-based science instruction [CRISI] to exemplary CRISI) and allows for the collection of data essential for critical teacher self-reflection (Howard, 2003), goal setting, and, inquiry into practice. The applicability of the CRISP framework spans all levels (K–graduate), but its creation, use, and evaluation are of paramount importance for K–12 African American students in science who deserve access to high-quality, equitable science learning experiences. The purpose of using the framework to create an analytic observation tool is to generate a valuable instrument that could inform teacher practices. Creating the CRISP framework is the first step to creating a developmental analytic observation tool. Such a tool and the associated framework can be used by school leaders to monitor and evaluate teachers as they engage in continued professional development related to CRP, IbSI, or CRISP. The framework should not take the place of formal teacher evaluation tools as it may not capture every aspect of instruction (e.g., planning) that is considered during formal evaluation.

Implications for STEM Teaching and Teacher Education

An exciting feature or outcome of creating this framework and the tool is that teachers and researchers can use it as an inquiry tool. Like other instruments, it can serve as a tool for data collection, and when used over time the teachers and researchers can chart professional growth. Those having their practice evaluated using the framework could also engage in targeted professional development in specific areas that require marked improvement (i.e., assessment—using various assessment strategies). For preservice teachers who are striving to be culturally relevant, they can use the framework/tool to investigate their practices during their internship or student teaching experience.

Furthermore, what is more promising is the impact such a framework, and subsequent tool, could have on the science learning and engagement of African American children. Taken separately, both CRP and IbSI have led to positive outcomes for students in general and science in particular. Putting the two together creates a synergy that could potentially positively influence African American students' success in science and their postsecondary degree and career choices. Finally, building and implementing the framework and tool serves as a challenge to the Western science status quo of excluding African American students whose home culture may differ from that of Western science. The dispositional nature of CRP allows teachers to critically self-reflect (Howard, 2003) to question and change their beliefs about African American students, and subsequently, change their practices in ways that create more science learning opportunities for African American students.

References

Aikenhead, G. S. (1996). Science education: Border crossing into the subculture of science. *Studies in Science Education, 27*(1), 1–52. doi:10.1080/03057269608560077

Aronson, B., & Laughter, J. (2016). The theory and practice of culturally relevant education: A synthesis of research across content areas. *Review of Educational Research, 86*(1), 163–206.

Bryan, L. A., & Atwater, M. M. (2002). Teacher beliefs and cultural models: A challenge for science teacher preparation programs. *Science Education, 86*(6), 821–839.

Freire, P. (1998). *Teachers as cultural workers. letters to those who dare teach. The edge: Critical studies in educational theory*. Boulder, CO: Westview Press.

Gay, G. (2010). *Culturally responsive teaching: Theory, practice and research*. New York, NY: Teachers College Press.

Gutstein, E., Lipman, P., Hernandez, P., & de los Reyes, R. (1997). Culturally relevant mathematics teaching in a Mexican American context. *Journal for Research in Mathematics Education, 28*(6), 709–737. doi:10.2307/749639

Howard, T. C. (2003). Culturally relevant pedagogy: Ingredients for critical teacher reflection. *Theory into Practice, 42*(3), 195–202. doi:10.1207/s15430421tip4203_5

Jegede, O. J., & Aikenhead, G. S. (1999). Transcending cultural borders: Implications for science teaching. *Research in Science & Technological Education, 17*(1), 45–66. doi:10.1080/02635149901701014

Kahle, J. B., Meece, J., & Scantlebury, K. (2000). Urban African-American middle school science students: Does standards-based teaching make a difference? *Journal of Research in Science Teaching, 37*(9), 1019–1041.

Ladson-Billings, G. (1990). Like lightning in a bottle: Attempting to capture the pedagogical excellence of successful teachers of Black students. *International Journal of Qualitative Studies in Education, 3*(4), 335–344.

Ladson-Billings, G. (1995). Toward a theory of culturally relevant pedagogy. *American Educational Research Journal, 32*(3), 465–491.

Ladson-Billings, G. (2001). *Crossing over to Canaan: The journey of new teachers in diverse classrooms*. San Francisco, CA: Jossey-Bass, Inc.

Ladson-Billings, G. (2006). Yes, but how do we do it? Practicing culturally relevant pedagogy. In J. Landsman & C. W. Lewis (Eds.), *White teachers/diverse classrooms: A guide to building inclusive schools, promoting high expectations, and eliminating racism* (pp. 29–42). Sterling, VA: Stylus.

Ladson-Billings, G. (2009). *The dreamkeepers: Successful teachers of African American children*. San Francisco, CA: Jossey-Bass.

Lopez, A. E. (2011). Culturally relevant pedagogy and critical literacy in diverse English classrooms: A case study of a secondary English teacher's activism and agency. *English Teaching: Practice and Critique, 10*(4), 75–93.

Marshall, J. C., Horton, R., & White, C. (2009). Equipping teachers. *The Science Teacher, 76*(4), 46.

Marshall, J. C., Smart, J. B., & Alston, D. M. (2017). Inquiry-based instruction: A possible solution to improving student learning of both science concepts and scientific practices. *International Journal of Science and Mathematics Education, 15*(5), 777–796.

Marshall, J. C., Smart, J. B, & Horton, R. M. (2010). The design and validation of EQUIP: An instrument to assess inquiry-based instruction. *International Journal of Science and Mathematics Education, 8*(2), 299–321.

Marx, R. W., Blumenfeld, P. C., Krajcik, J. S., Fishman, B., Soloway, E., Geier, R., & Tal, R. T. (2004). Inquiry-based science in the middle grades: Assessment of learning in urban systemic reform. *Journal of Research in Science Teaching, 41*(10), 1063–1080. doi:10.1002/tea/20039

National Research Council. (1996). *National science education standards*. Washington, DC: National Academies Press.

Parsons, E. C., Travis, C., & Smith-Simpson, J. (2005). The black cultural ethos, students' instructional context preferences and student achievement: An examination of culturally congruent science instruction in the eighth grade classes of one African American and on Euro-America teacher. *The Journal of Negro Education Review, 56*(2), 183–203.

Popham, W. J. (2017). *Classroom assessment: What teachers need to know* (8th ed.). Boston, MA: Pearson Education, Inc.

Prime, G. M., & Miranda, R. J. (2006). Urban public high school teachers' beliefs about science learner characteristics: Implications for curriculum. *Urban Education, 41*(5), 506–532.

Prince, M. (2004). Does active learning work? A review of the research. *Journal of Engineering Education, 93*(3), 223–231.

Sawada, D., Piburn, M. D., Judson, E., Turley, J., Falconer, K., Benford, R., & Bloom, I. (2002). Measuring reform practices in science and mathematics classrooms: The reformed teaching observation protocol. *School Science and Mathematics, 102*(6), 245–253.

Ware, F. (2006). Warm demander pedagogy: Culturally responsive teaching that supports a culture of achievement for African American students. *Urban Education, 41*(4), 427–456.

Antiracist Curriculum and Pedagogies in Science Teacher Education

FELICIA MOORE MENSAH

Abstract
The education of African American children cannot be lost or neglected. Consequently, we need teachers to understand the impact of race and racism on education by engaging them in conversations about race and racism and creating the support for all teachers to develop a language to do so. This chapter speaks to elements of theory, practice, and research with emphasis on the preparation of science teachers to engage in discussions of race and the development of anti-racist curriculum where race is foregrounded in science teacher education and science classrooms.

Introduction

At a locally held conference, a mathematics colleague and I co-facilitated a working group with science and mathematics teachers on the subject of teacher education. From the discussions, many issues emerged as we talked about practices to support science and mathematics teachers who teach African American students. Many familiar challenges were discussed. These challenges, we noted, are often left to be addressed in teacher education and teacher professional development, which may be unfair or unrealistic. One challenge derives from what Ladson-Billings

(2007) calls an educational debt which cannot be addressed solely by attending to content and pedagogy. Another challenge is developing the ability to teach for diversity, and this involves helping teachers to acquire the dispositions, skills, knowledge, and a broad understanding of diversity to teach all students (Moore, 2008a). Furthermore, "preservice teachers who have few experiences with diversity acknowledge that they need more opportunities to learn about differences and to make multicultural connections within science teaching and learning" (Moore, 2008a, p. 104). Likewise, novice teachers report that they want more from their teacher education programs to prepare them for teaching in culturally diverse classrooms (Colón-Muñiz, Brady, & SooHoo, 2010). As I think about the participants from that conference and the graduate students I teach, both practicing and prospective teachers, it is necessary to engage in conversations of race and racism with science and mathematics teachers. For me, this work is situated in science teacher education.

Teacher education programs, for the most part, do not provide opportunities for teacher candidates to have conversations about race and racism. Sometimes the barriers to doing so are due to intentional neglect, but mainly as a result of implicit messaging (Picower, 2009). In such settings, discussions of culturally relevant teaching (Ladson-Billings, 2014) or culturally responsive teaching (Gay, 2000), and other inclusive pedagogical approaches may be easier to discuss than race-based or antiracist pedagogies. I make an argument here that I have made about culturally relevant science teaching: If we want our teachers to teach in culturally relevant ways, then we must allow them to learn in this way (Mensah, 2011). Similarly, if we want our teachers to understand or discuss how race impacts teaching and learning of STEM-CS (Science, Technology, Engineering, Mathematics, and Computer Science), then we have to engage them in conversations about race and racism and elicit their thoughts about teaching African American students. This would involve making discussions of race a necessary component in science teacher education.

Hence, this chapter speaks to elements of theory, practice, and research with emphasis on the preparation of science teachers to engage in discussions of race, and the development of antiracist curricula where race is foregrounded in science teacher education. Using the method of retrospective coding (Mensah, 2013), I offer quotations from preservice teachers engaged in race conversations intended to expand their ideas of race and its influence on science education. These excerpts come from course assignments over several years that show the trajectory of my work with teacher candidates where I start with diversity and move toward race as the topic for conversation in my courses. I argue for race-visible pedagogies and antiracist curriculum as one way to engage all teachers—including African American teachers—to build their knowledge to address race and racism in their teaching of science in K–16 settings.

Reappearance of Race in Education

Peller (1990) acknowledged from the work of critical race scholars of the law that race consciousness has taken on a "reappearance and refinement" in the literature (p. 258) due to explicit conversations that once had been considered "taboo…within mainstream American politics for far longer within particular conventions of law and legal scholarship" (p. 258). In the current racial and political climate of the United States (US) and the world, people are at least becoming more conscious of race issues, whether through killings of Black men, attacks on women of color, physical violence of children in schools, and political musings against people of cultural and religious difference. People in many ways are forced to deal with the US's avoidance of race and racism that has extended to educational settings. However, Martin (2013) stated, "Discussions of culture, more than race and racism, predominate" (p. 320). Culture may feel easier to talk about over race and racism because the controversial topic of race causes many students and teachers to feel anxious (Richeson & Shelton, 2007; Vorauer, Hunter, Main, & Roy, 2000), or they develop strategies that substitute for discussions of race as culture (Picower, 2009), and they try not to acknowledge race by using colorblind language (Marx, 2006). Still, others avoid the topic altogether (Goff,

Steele, & Davies, 2008). In addition to this, several scholars discuss the tactics people take to avoid discussions of race. For example, Apfelbaum, Sommers, and Norton (2008) refer to avoiding the topic of race as "strategic colorblindness" and suggest that it stems from White peoples' concern with appearing prejudiced or racist. Teaching approaches are discussed devoid of race or any acknowledgment of how race and racism affect educational equity for students and for teachers and their growth as educators.

Even less is done to address discussions of race in science teacher education (exceptions, Alexakos et al., 2016; Mensah, 2016; Mensah & Jackson, 2018; Sheth, 2018). If conversations about the impact of race and racism in science education are not happening, and if science teachers continue to avoid these discussions, then the educational debt, alluded to earlier, will persist. Elevating educational achievement for African American children and their participation in science will not be realized either. Therefore, approaches to address bias and to start conversations about race are needed in science teacher education.

Most teachers will likely have students in their classrooms who will come from diverse racial, ethnic, linguistic, and religious groups during their careers, especially more now than in the past due to increasing numbers of students from diverse cultural backgrounds, globalization and immigration, and bi- and multi-racial children from interracial group relationships. Experiencing greater diversity will be true for teachers across rural, suburban, and urban schools in the United States as well as in global contexts (Banks & Banks, 2010). When Ladson-Billings (2007) speaks of an "education debt," she holds everyone accountable for the education of all children and states, "we have accumulated this problem as a result of centuries of neglect and denial of education to entire groups of students" (p. 321). When we think about the neglect of entire groups of students in classrooms, the failure to center race in the science education of African American students and teachers is a serious omission that has to be addressed (Mutegi, 2011; Parsons, 2014).

I have started this work in my own teaching and preparation of pre-service teachers and with my doctoral students, many of whom are practicing teachers, and others who are planning to become teacher educators. The focus of this work is the design and implementation of race-based pedagogies that promote an understanding of the impact of race and racism on the teaching and learning of African American students in science, and the pedagogical approaches needed to disrupt the persistent inequities seen in science education.

Starting Conversations About Race

In a White-dominated society, most of us have received little or no education and information about racism and are thus unprepared to think or to talk about it critically (DiAngelo, 2006; Matias, 2013). Of course, we have personal experiences; however, this is not enough to contend with the years of explicit and implicit bias that have contributed to our misunderstandings of self and others. We are not immune to what society has forced on us. We are not exempt from the many influences of race and racism on our thinking. Hence, we must engage in conversations about race explicitly. For teachers, these conversations about race and their views about education and the students they teach, mainly students of color, have to also ignite critical reflection on their practice and how they engage students in learning. Discussions of race in teacher education can begin to address the deeper meanings held by teacher candidates and the influence that race and racism have on educational equity.

In previous work in science teacher education, I have used many approaches to address diversity and equity in science education. These approaches reveal assumptions and biases teachers have about diversity and teaching of students of diverse cultural, ethnic, linguistic, and racial backgrounds. One approach has been the Book Club discussions which have been useful for individual, collaborative, and collective learning (Mensah, 2009; Moore, 2008a, 2008b). In this approach, teacher candidates read multicultural literature in a science methods course. This exposure to multiple perspectives across contexts created space within the teacher education curriculum to confront assumptions and beliefs through intimate conversations about issues of diversity and equity. The conversations in many ways "forced" the pre-service teachers to reveal the underlying assumptions and biases they held about teaching science and teaching students of color (Mensah, 2009). The pre-service teachers were also placed in partnership elementary schools which served predominately African American student populations. In the classrooms, the pre-service teachers work with children and teach science.

The findings from the Book Club covered five major themes: (a) Relevancy, using a multicultural text in a science methods course; (b) Revelation, revealing assumptions and biases about issues of diversity and teaching science; (c) Responsiveness, forcing a response to issues of diversity in science education; (d) Reflection, developing critical and reflective science teachers; and (e) Reformation, gaining a deeper understanding of diversity by changing ideological beliefs. These five themes suggest that the overall structure and theoretical foundation of the Book Club promoted teacher learning about complex issues in urban science education for diversity and equity. For example, the Book Club challenged deficit models of thinking and teaching, and at the same time, established a purpose for educating critically reflective teachers to improve science education in urban schools. One pre-service teacher wrote of her learning and participating in her Book Club group:

> Our book club helped to shed light on my flawed, and almost, dismissive approach to this book, and how this could be incredibly detrimental in the classroom. Accordingly, overtime, I realized how cognizant I need to be of my existing biases and sensitive to the often subconscious effect of my ethnocentricity. Stereotypes are easy; just as I needed to focus to find the real value and meaning in the text, I could find constructive insight and realize a mutually beneficial experience/interaction from a deliberate effort to understand my students and their backgrounds. (Pre-service Teacher, Final BC Group Reflections)

The pre-service teachers also discussed in their groups ways to design science curriculum that considered the students they were teaching. For instance, one pre-service teacher stated on her final Book Club paper, *"We then need to adjust our curriculum to benefit the students from all backgrounds"* and another shared, *"The book helped me to understand that curriculum needs to be tailored to suit the needs of the students in order for it to be the most effective."* Though the learning from the Book Club brought out many relevant ideas about equity and diversity, explicit language about race was not so evident. Still, the Book Club was successful in allowing the pre-service teachers to reveal beliefs and assumptions about issues of diversity in science, to reflect on their role as teachers, and to gain a deeper understanding, awareness, and sensitivity to issues of diversity in the science education of African American students. For them, "issues of diversity [become] central rather than peripheral" (Villegas & Lucas, 2002, p. 21), and the goal was to better understand their practice when teaching students of color, teaching in diverse classroom settings, and teaching science.

From this early work with pre-service teachers, revealing assumptions and biases is a first step in transforming teacher practices to address race in science education. As teachers begin to understand race and racism within society, how it impacts their teaching and relationships with students, a more deliberate attempt was made to foreground discussions of race and racism in teacher education. There is a need to transform not only how teachers think about these issues, but also how to get teachers to critique their own teaching practices in the development of language to discuss race and racism in teacher education. These conversations ultimately build greater awareness of racial inequity in education.

Teaching Racial Literacy

As mentioned earlier, if we want teachers to understand the impact of race and racism on education and on their teaching, then we have to engage them in conversations about race and racism while supporting them in the development of language to do so. As scholars note, work in teacher education that involves conversations about race is needed (Mensah, 2015; Milner, 2010). Emergent research in the area of conversations about race in teacher education is trending (Alexakos *et al.*, 2016; Buchanan, 2015; Coles-Ritchie & Smith, 2017; Mensah, 2015; Mensah & Jackson, 2018; Singleton, 2015). As examples, Alexakos and his colleagues report the use of personal narratives and a heuristic to assist teachers in discussing "thorny issues" such as gender and race in science teacher education. They suggested possible solutions and interventions to create "safe, supporting, and healing spaces for complex, often emotionally painful discussions" (p. 741). Buchanan's work with White pre-service teachers and their understanding of race involved efforts to intentionally begin conversations using questionnaires, online discussions, and individual written reflections. These activities allowed the pre-service teachers to identify and articulate their beliefs while recognizing others' beliefs. Coles-Ritchie and Smith in their study explored how African American, Polynesian, and White in-service teachers who participated in "Courageous Conversations" (Singleton, 2015) professional development addressed or avoided talking about race in their elementary schools. From conducting in-depth interviews, their findings indicated that to have equity in schools more professional development about race talk in elementary schools is needed. All of these studies support the notion that conversations about race can be an explicit part of the teacher education curriculum and teacher professional development goals.

Based upon my previous work with elementary pre-service teachers, I specifically designed a course to engage in conversations about race in teacher education (Mensah, 2015). Utilizing a range of pedagogical approaches, all centered on the development of racial literacy (Sealey-Ruiz, 2011; Stevenson, 2013), the purpose of the course was to assist advanced master's students and doctoral students to develop racial literacy. Racial literacy, defined by Sealy-Ruiz, is "the ability to discuss the implications of race and American racism in edifying and constructive ways" (2011, p. 25). Twine and Steinbugler (2006) define racial literacy as "a way of perceiving and responding to the racial climate and racial structures individuals encounter" (p. 344). Developing racial literacy is an ongoing process that requires self-education with an emphasis on learning in an interactive way rather than knowing (Guinier, 2004; Rogers & Mosley, 2006; Twine & Steinbugler, 2006). For the students in the course, developing racial literacy was a primary goal. In the next section, I highlight a few instances of how graduate students in the course began developing racial literacy and how understandings of race in their teacher education preparation may support further learning and their ability to make changes in curriculum and pedagogy to support students of color.

Learning to Develop Racial Literacy

There were varied pedagogical practices that were created for the course. They were reading book chapters and published articles on topics of race, racism, and equity in teacher education, engaging in class activities focused on race and racism, watching video and movies clips that raised issues of race and racism, and co-teaching with opportunities for students to co-lead a class session. In addition, personal journaling was used in the course for students to develop their language, reflect on their own identities, teaching practices, and conversations in class. The journal was also used for them to talk about race from connecting with the course readings, activities, and other pedagogical practices used in the course. For example, from the video clips and conversations in class, one student became more conscious of how others viewed her. From reflecting on past interactions, she realized that racism may show up in "subtle" ways:

> The videos and discussions also increased my awareness of racism against me personally. I never really thought about others being racist towards me but thinking deeply about certain situations has made me realize that subtle events have occurred in the past that would be considered racist. (In-service Teacher, Final Thoughts Paper)

The students also participated in developing critical thinking and conversation skills around issues and topics involving racism, discrimination, and prejudice in the context of their education as graduate students and as New York City (NYC) public school teachers. For example, a student, also a high school teacher, remarked how race and racism are "present" yet "unspoken":

> For me, race/racism is the prevalent unspoken topic in teacher education today. It is an issue present in every classroom in NYC, across the US and the majority of the world. Yet the recent push for educational reforms fails

to mention this issue in any way, shape or form. (In-service Teacher, Final Thoughts Paper)

Furthermore, the conversations that centered race and racism also revealed how important it was to discuss this in the content areas. The course enrolled graduate students across multiple programs in the college. One graduate student not in science education mentioned how he became more aware of science as a field in relation to our discussions of race across multiple content areas, including the teaching of science:

> I think the incorporation of race [in the course] certainly widened my research interests. Furthermore, I never really knew how strongly race played a role in science education. Although this is not my initial field, I was very interested to see how educators and participants conceptualized science class relying on dated stereotypes that are perpetuated by current institutions and required curricula. (In-service Teacher, Course Evaluation)

Therefore, the discussions, readings, and other instructional strategies used in the course assisted the teachers in gaining greater understanding of how to connect race to larger issues in teacher education and society, and to their teaching of science and other subject areas, and themselves. Developing racial literacy involves engaging in constructive conversations of race and racism that begin with creating opportunities for these conversations to occur. Furthermore, it requires deliberate attention to be explicit in how to talk about race and racism in science teacher education. For an explicit focus on race and racism in science teacher education, I argue for antiracist curriculum and pedagogies that center race in science classrooms. This requires teachers to think more intentionally about what is being taught and how it is being taught in science classrooms.

Racist Curriculum and Pedagogies

Essentially, race covers all facets of educational life from curriculum (Banks & Banks, 2010), to school policies (Oakes, 2005), to achievement rates (Ladson-Billings, 2007), to particular practices and school structures (Leonardo & Grubb, 2014), and to our living, learning, and teaching experiences (Hughes & Berry, 2012). Even so, on a daily basis in our classrooms, White children are taught that they are White. This comes through in the curriculum that prioritizes Western ways of knowing and Eurocentric perspectives, and in classroom practices that incorporate their mainstream experiences and interactions over others. While doing this, teachers also create and maintain an educational distance where African American students—and other students of color—are not engaged, and learning is not connected to their experiences and histories. MacDonald (2003) contends that curriculum reforms "have failed to override the influences of what students bring to the school—their neighborhood, socio-economic status, gender, and ethnicity" and that "schooling, and, thereby, the school curriculum, is generally reproductive. It is active in reproducing the economic and cultural imbalances upon which society is built" (p. 145), namely the White dominant views of teaching, learning, and living.

Further, curriculum has the ability to recreate race and racial inequity because students can read their existence or nonexistence in the textbooks and curricular materials used in the classroom. Leonardo and Grubb (2014) stated, "The curriculum project becomes a racial project" (p. 18) because the traditional, White-normed curriculum is not based on the experiences of African American students and students from diverse cultural and linguistic backgrounds. The curriculum taught in schools noticeably does not attend to the particular experiences and knowledges of students of color. Consequently, African American students feel excluded and marginalized in many K–12 to college science classrooms (Beasley & Fischer, 2012; Mensah & Jackson, 2018; Mutegi, 2011). These remarks communicate the critique of traditional, White-normed curriculum as not meeting the needs of African American students. There is a need to be intentional and inclusive in developing curriculum that addresses issues of race and racism in science education.

Antiracist Curriculum and Pedagogies in Science Education

Antiracist curriculum and antiracist pedagogies are grounded in critical theories of education and multicultural education. Antiracist pedagogy has been called liberatory, critical, or radical pedagogy. It is rooted in social justice perspectives and is dedicated to education reform by exposing racialized power relations (Klassen & Carr, 1996). For example, Blakeney (2005) defined antiracist pedagogy to include "explicit instruction on confronting racism without reservation or risk of ostracism, both of which are necessary in a society that mandates the purpose of public education as the production of democratic citizenry" and "aims at transformation by challenging the individual as well as the structural system that perpetuates racism" (p. 120).

There are a few ways that researchers have categorized levels of antiracist curriculum and teaching. For instance, Louis Derman-Sparks and Carol Phillips (1997) conceptualize antiracist multicultural education as four different levels: a) *the basic level* involves teachers engaging in a single event or activity; b) *the project or unit approach* involves inserting into the curriculum substantive cultural content; c) *the integrated or transformative level* involves integrating multicultural content across all subject areas; and d) *the social action level* encourages students to act for social justice. Similarly, Banks and Banks (2010) outline four levels of multicultural curriculum reform that are akin to the categories of Derman-Sparks and Phillips. Their four levels are: a) *the contributions approach* with teaching the heroes, holidays, and discrete cultural elements; b) *the additive approach* with teaching of cultural content, concepts, and themes added into the curriculum with no disruption of the Eurocentric or mainstream canon; c) *the transformative approach* with teaching that restructures the curriculum to include different perspectives; and d) *the social action approach* with teaching that extends the transformative approach with action on important social issues and goals of social justice.

Both approaches may be viewed as occurring on a continuum for developing curriculum and educating teachers toward antiracist pedagogy and teaching (Mensah et al., 2018). However, the lower levels—that is, the basic and the project or unit approach for Derman-Sparks and Phillips (1997) and the contributions and additive approach for Banks and Banks (2010)—involve teachers engaging in a single event or activity or introducing an example of cultural diversity into their curriculum. These basic levels are problematic because they frequently lead to an increase in stereotypes about a particular cultural group. The aim should be the higher levels—that is, the integrated or transformative and the social action approaches. At these levels, there is substantive change with a deliberate regard to culture, diversity, and multiple perspectives in the existing curriculum or the creation of new curriculum. The experiences of marginalized groups are included, and all students receive a more "expansive 'truth'" (Leonardo & Grubb, 2014, p. 21) of the society and the roles that people have contributed. At the higher levels, learning about multiple perspectives and what this means for various ethnic groups, power relationships in society, and how knowledge is constructed are addressed. Teachers encourage students to be critical of knowledge, consider how knowledge is constructed, and discuss the association of knowledge to issues of race and power.

In addition, the upper levels involve integrating multicultural content throughout all subject areas. This will also mean including discussions of race across multiple subject areas as well. For example, incorporating the Tuskegee Syphilis Study into a science curriculum covers various subject areas and invites a discussion of race and racism. In brief, discussions may involve the African American men and the participation in the study for science, health and disease, and broader fields of study, such as US history and the particular time the Tuskegee Study was done, along with ethics and human rights. For English connections, doing research on personal narratives, documentaries, and newspaper clippings where multiple voices are read and heard. Broader discussions and questions are asked to cover the sociocultural, sociogeographical, sociohistorical, and sociopolitical contexts for deeper understandings. A social justice project may be completed where students hold a Science and Health Fair for their community. Here, they may produce brochures to educate people on their rights as patients or create a Patient Bill of Rights that is easy to read and understand. The brochure can include questions patients may ask when going to the doctor or seeking health services. Therefore, at the higher levels, a curriculum centered on the history and experiences of African Americans provides a strong connection to understanding the impact of race and racism on people of color, our society, and education for what is taught and learned in school. "Framing the curriculum regulates how race history unfolds by giving it shape, providing its sequence of events, and stamping it with a sense of time" (Leonardo & Grubb, 2014, p. 23). Antiracist curriculum and pedagogies in science education make explicit the necessity to address the impact of race on teaching and learning.

Returning to the Purpose and Conclusion

We cannot neglect or overlook the education of African American children. The opportunities to talk about issues of race and racism in science teacher education can be created by giving explicit attention to assisting all teachers in developing racial literacy and antiracist curriculum and pedagogies. The development of science curriculum or science lessons and unit plans is a common activity for teacher candidates and practicing teachers. Race-based curriculum and pedagogies can be supported in teacher education, but this must be sustained with ongoing professional development to challenge and reform existing curriculum to be more inclusive of race conversations.

For antiracist curriculum and pedagogies to work, science educators, and STEM-CS educators as well, have to re-consider frameworks that will support teachers in reflecting critically on their practices to center race. The upper levels for antiracist curriculum and pedagogies from Derman-Sparks and Phillips (1997) or Banks and Banks (2010) are valuable guides for consideration (Mensah, 2011). The frameworks require antiracist pedagogies that also include engaging in conversations to support reform and restructuring of teacher education programs. Recently in my own work, I am giving even more thought and attention to focus on antiracist curriculum and pedagogies in teacher education curriculum for both pre-service and in-service teachers by fostering an environment that promotes the development of racial literacy and the centering of race in science teacher education (Mensah, 2015; Mensah et al., 2018; Mensah & Jackson, 2018). The goal is the transformation of the curriculum and pedagogies in ways that lead to change in teachers' practices and their views of science and education and the teaching of science to students of color.

Because our African American students have suffered so much and for so long from educational inequities due to our failure to center their experiences, we have to be better at supporting their educational needs through the work we do with their teachers. Antiracist curriculum and pedagogies certainly involve intentional effort; we have to believe our children are more than deserving of an education and curriculum that speaks to their needs and experiences, a curriculum that highlights their culture and language, a curriculum that acknowledges their histories and contributions, and a curriculum that engages them to move toward action for the betterment of themselves and their communities. Teacher education is one place to contribute to supporting African American learners in the science classroom by supporting teachers to make classroom environments pedagogically engaging and socially supportive for all students, and where race is not peripheral to teaching and learning but a direct and central place in the classroom.

References

Alexakos, K., Pride, L., Amat, A., Tsetsakos, P., Lee, K. J., Paylor Smith, C., Zapata, C., Wright, S., & Smith, T. (2016). Mindfulness and discussing "thorny" issues in the classroom. *Cultural Studies of Science Education*, 11(3), 741–769.

Apfelbaum, E. P., Sommers, S. R., & Norton, M. I. (2008). Seeing race and seeming racist? Evaluating strategic colorblindness in social interaction. *Journal of Personality and Social Psychology*, 95(4), 918–932.

Banks, J. A., & Banks, C. M. (2010). *Multicultural education: Issues and perspectives* (7th ed.). Hoboken, NJ: John Wiley & Sons Publishers.

Beasley, M. A., & Fischer, M. J. (2012). Why they leave: The impact of stereotype threat on the attrition of women and minorities from science, math and engineering majors. *Social Psychology of Education*, 15(4), 427–448.

Blakeney, A. M. (2005). Antiracist pedagogy: Definition, theory, and professional development. *Journal of Curriculum and Pedagogy*, 2(1), 119–132.

Buchanan, L. B. (2015). "We make it controversial": Elementary preservice teachers' beliefs about race. *Teacher Education Quarterly*, 42(1), 3–26.

Coles-Ritchie, M., & Smith, R. R. (2017). Taking the risk to engage in race talk: Professional development in elementary schools. *International Journal of Inclusive Education*, 21(2), 172–186.

Colón-Muñiz, A., Brady, J., & SooHoo, S. (2010). What do graduates say about multicultural teacher education? *Issues in Teacher Education, Spring*, 85–108.

Derman-Sparks, L., & Phillips, B. C. (1997). *Teaching/learning anti-racism*. New York, NY: Teachers College Press.

Diangelo, R. (2006). The production of whiteness in education: Asian international students in a college classroom. *Teachers College Record, 108*(10), 1983–2000.

Gay, G. (2000). *Culturally responsive teaching: Theory, research, and practice*. New York, NY: Teachers College Press.

Goff, P. A., Steele, C. M., & Davies, P. G. (2008). The space between us: Stereotype threat and distance in interracial contexts. *Journal of Personality and Social Psychology, 94*, 91–107. doi:10.1037/0022–3514.94.1.91

Guinier, L. (2004). From racial liberalism to racial literacy: Brown v. Board of Education and the interest-divergence dilemma. *Journal of American History, 91*(1), 92–118.

Hughes, S., & Berry, T. R. (2012). *The evolving significance of race: Living, learning, & teaching*. New York, NY: Peter Lang.

Klassen, T. R., & Carr, P. R. (1996). The role of racial minority teachers in anti-racist education. *Canadian Ethnic Studies, 28*(2), 126–138.

Ladson-Billings, G. (2014). Culturally relevant pedagogy 2.0: aka the remix. Harvard Educational Review, 84(1), 74–84.

Ladson-Billings, G. (2007). Pushing past the achievement gap: An essay on the language of deficit. *The Journal of Negro Education, 76*(3), 316–323.

Leonardo, Z., & Grubb, W. N. (2014). *Education and racism: A primer on issues and dilemmas*. New York, NY: Routledge.

MacDonald, D. (2003). Curriculum change and the post-modern world: Is the school curriculum-reform movement an anachronism? *Journal of Curriculum Studies, 35*(2), 139–149.

Martin, D. B. (2013). Race, racial projects, and mathematics education. *Journal for Research in Mathematics Education, 44*(1), 316–333.

Marx, S. (2006). *Revealing the invisible: Passive racism in teacher education*. New York, NY: Routledge.

Matias, C. E. (2013). Tears worth telling: Urban teaching and the possibilities of racial justice. *Multicultural Perspectives, 15*(4), 187–193.

Mensah, F. M. (2009). Confronting assumptions, biases, and stereotypes in preservice teachers' conceptualizations of science teaching through the use of book club. *Journal of Research in Science Teaching, 46*(9), 1041–1066.

Mensah, F. M. (2011). A case for culturally relevant teaching in science education and lesson learned for teacher education. *The Journal of Negro Education, 80*(3), 296–309.

Mensah, F. M. (2013). Theoretically and practically speaking, what is needed in diversity and equity in science teaching and learning? *Theory into Practice, 52*(1), 66–72. doi: 10.1080/00405841.2013.743781

Mensah, F. M. (2015, April). When race becomes a focus in teacher education. AERA Annual International Conference, Chicago, IL.

Mensah, F. M. (2016). Positional identity as a framework to studying science teacher identity: Looking at the experiences of teachers of color. In L. Avraamidou (Ed.), *Studying science teacher identity: Theoretical perspectives, methodological approaches and empirical findings* (pp. 49–69). The Netherlands: Sense Publishers.

Mensah, F. M., Brown, J. C., Titu, P., Rozowa, P., Sivaraj, R., & Heydari, R. (2018): Preservice and inservice teachers' ideas of multiculturalism: Explorations across two science methods courses in two different contexts. *Journal of Science Teacher Education, 29*(2), 128–147. doi:1046560X.2018.1425820

Mensah, F. M., & Jackson, I. (2018). Whiteness as property in science teacher education. *Teachers College Record, 120*(1), 1–38. Retrieved from http://www.tcrecord.org. ID Number: 21958.

Milner, H. R. (2010). What does teacher education have to do with teaching? Implications for diversity studies. *Journal of Teacher Education, 61*(1–2), 118–131.

Moore, F. M. (2008a). Preparing preservice teachers for urban elementary science classrooms: Challenging cultural biases toward diverse students. *Journal of Science Teacher Education, 19*(1), 85–109.

Moore, F. M. (2008b). The role of the elementary science teacher and linguistic diversity. *Journal of Elementary Science Education, 20*(3), 49–61.

Mutegi, J. W. (2011). The inadequacies of "Science for All" and the necessity and nature of a socially transformative curriculum approach for African American science education. *Journal of Research in Science Teaching, 48*(3), 301–316.

Oakes, J. (2005). *Keeping track.* New Haven, CT: Yale University Press.

Parsons, E. C. (2014). Unpacking and critically synthesizing the literature on race and ethnicity in science education. In N. G. Lederman & S. K. Abell (Eds.), *Handbook of research in science education* (pp. 687–767). New York, NY: Routledge.

Peller, G. (1990). Race consciousness. *Duke Law Journal, 1990*, 758–847.

Picower, B. (2009). The unexamined whiteness of teaching: How white teachers maintain and enact dominant racial ideologies. *Race Ethnicity and Education, 12*(2), 197–215.

Richeson, J. A., & Shelton, J. N. (2007). Negotiating interracial interactions: Costs, consequences, and possibilities. *Current Directions in Psychological Science, 16*(6), 316–320.

Rogers, R., & Mosley, M. (2006). Racial literacy in a second-grade classroom: Critical race theory, whiteness studies, and literacy research. *Reading Research Quarterly, 41*(4), 462–495.

Sealey-Ruiz, Y. (2011). Learning to talk and write about race: Developing racial literacy in a college English classroom. *English Quarterly Canada, 42*(1/2), 24.

Sheth, J. J. (2018). Grappling with racism as foundational practice of science teaching. *Science Education*, 1–24. doi:10.1002/sce.21450

Singleton, G. E. (2015). *Courageous conversations about race: A field guide for achieving equity in schools* (2nd ed.). Thousand Oaks, CA: Sage.

Stevenson, H. (2013). *Promoting racial literacy in schools: Differences that make a difference* [Kindle Android version]. Retrieved from Amazon.com

Twine, F. W., & Steinbugler, A. C. (2006). The gap between whites and whiteness: Interracial intimacy and racial literacy. *Du Bois Review: Social Science Research on Race, 3*(2), 341–363.

Villegas, A. M., & Lucas, T. (2002). Preparing culturally responsive teachers: Rethinking the curriculum. *Journal of Teacher Education, 53*(1), 20–32.

Vorauer, J. D., Hunter, A. J., Main, K. J., & Roy, S. A. (2000). Meta-stereotype activation: Evidence from indirect measures for specific evaluative concerns experienced by members of dominant groups in intergroup interaction. *Journal of Personality and Social Psychology, 78*(4), 690–707.

Contributors

Keisha McIntosh Allen, Ed.D. is an assistant professor of education in the Secondary Education Program at University of Maryland, Baltimore County. Her research, teaching, and service focus on preservice and in-service teachers' critical and asset pedagogies within in-school and out-of-school contexts.

Elsa Bailey, PhD, is the principal researcher/director of *Elsa Bailey Consulting*. Based in the San Francisco Bay area in California, her professional activities reach across informal and formal education projects, with a concentration on STEM, social justice, informal learning, and early childhood learning.

Scott A. Chamberlin, PhD, is professor of mathematics in the School of Teacher Education at the University of Wyoming. He has served a co-principal investigator on the Wyoming Interns to Teacher Scholars (WITS), Robert Noyce grant. His research interests include teacher education and gifted education.

Julius Davis, Ed.D. is an associate professor of mathematics education in the Department of Teaching, Learning, and Professional Development in the College of Education at Bowie State University. Davis has two main strands of research focused on Black male students and teachers in urban schools.

Helen Douglass, PhD, is assistant professor of STEM education in the Henry Kendall College of Arts & Sciences at the University of Tulsa. Her research interests include gender equity in STEM education, formal and informal learning spaces, and equitable and inclusive STEM teaching and learning.

Roni Ellington, PhD, completed her doctorate in the Department of Curriculum and Instruction at the University of Maryland, College Park. She is currently associate professor of Mathematics Education at Morgan State University and is principal investigator of the NSF-funded SEMINAL project that focuses on integrating culturally responsive teaching in undergraduate mathematics instruction.

Ramon B. Goings, Ed.D. is an assistant professor of educational leadership at Loyola University Maryland. His research focuses on the academic and social experiences of gifted/high-achieving Black males PK–PhD, nontraditional students, diversifying the teacher and school leader workforce, and historically Black colleges and universities

Lesley K. Etienne earned his PhD from Antioch University. He works at Indiana University Purdue University, Indianapolis as a visiting professor in the School of Education and an instructor in the Africana Studies Program.

Jacqueline Leonard, PhD, completed her doctorate in the Department of Curriculum and Instruction at the University of Maryland, College Park. She is currently professor of Mathematics Education at the University of Wyoming and the 2018–19 Fulbright STEM Research Chair at the University of Calgary.

Felicia Moore Mensah (PhD, Florida State University) is associate dean and professor of Science Education at Teachers College, Columbia University. With several honors and awards, including ASTE Outstanding Science Teacher Educator of the Year and AERA Early Career Award, her research addresses issues of diversity and equity in teacher education.

Crystal H. Morton is an associate professor of Mathematics Education at Indiana University–Purdue University at Indianapolis (IUPUI). She is the founder and director of Girls STEM Institute. Her most recent publication is in the National Council of Teachers of Mathematics 2018 Annual Perspectives in Mathematics Education.

Jomo W. Mutegi earned his PhD from Florida State University. Currently, associate professor of science education at Indiana University, IUPUI and principal investigator of the $(ES)^2$ STEM Learning Lab, he has recent publications in the *Journal of Research in Science Teaching* and *Teachers College Record*.

Glenda Prime, PhD is currently chair of the Department that houses 9 graduate programs in Education at Morgan State University. She pursues two lines of research: doctoral education and social and cultural impacts on STEM education of African American Children.

Before receiving her PhD from the University of Pennsylvania, **Gale Seiler** was a high school science teacher for 17 years. She is now an associate professor at Iowa State University, where she teaches and does research in the areas of sociocultural approaches to science education.

Vanessa Dodo Seriki, PhD, received her doctorate in Education, with an emphasis in Science and Teacher Education, from the Ohio State University. She is currently an associate professor of Science Education at Morgan State University in Baltimore, Maryland.

Geeta Verma, PhD, is associate professor of science education at the University of Colorado Denver. She served as co-principal investigator on the Dinosaurs, Denver and Climate Change project. Her research focus is creating accessible and equitable STEM learning opportunities in formal and informal learning environments.

www.ingramcontent.com/pod-product-compliance
Ingram Content Group UK Ltd.
Pitfield, Milton Keynes, MK11 3LW, UK
UKHW021846140426
5217IPUK00022B/1610